Keeping the Last Best Fishery

advice from Montana's biologists to the next generation

Featuring
Amber Steed
Barry Hansen
Beth Gardner
Bob Gresswell
Brad Shepard
Brian Marotz
Caleb Bollman
Carter Kruse
Chris Clancy
Chris Hunter
David Brooks
Jim Dunnigan
Jim Vashro
Leslie Nyce
Matt Boyer
Mike Jakober
Mike Ruggles
Pat Clancey
Pat Saffel
Tom McMahon
Travis Horton
Wade Fredenberg

Edited by
Niall G. Clancy

Foreword by
Dick Vincent

Artwork by
T. David Ritter

All proceeds from the sale of this book go to the Montana Chapter of AFS' Resource Action Fund and Wally McClure Scholarship

Keeping the Last Best Fishery

advice from Montana's biologists to the next generation

edited by
Niall G. Clancy

Published on behalf of the
Montana Chapter of the American Fisheries Society
Montana, USA
2024

A suggested citation format for this book follows.

Entire Book

Clancy, N. G., editor. 2024. Keeping the last best fishery: advice from Montana's biologists to the next generation. Montana Chapter of the American Fisheries Society, Montana, USA.

Individual Interview

Shepard, B. 2024. Values and skills. Pages 6-9 *in* N. G. Clancy, editor. Keeping the last best fishery: advice from Montana's biologists to the next generation. Montana Chapter of the American Fisheries Society, Montana, USA.

Montana Chapter of the American Fisheries Society website: units.fisheries.org/montana/

"Know what the hell you're talking about."

-Dr. Robb Leary

when asked about his advice to the next generation
over beers at the 2019 MTAFS meeting in Billings

Dedicated to the scientific mentors of my early career (that aren't featured in this book) …

Tom Schmit
Biology Teacher,
Hamilton High School, Hamilton, MT

Marshall Bloom
Assoc. Director for Science Management,
Rocky Mountain Laboratories, Hamilton, MT

Wyatt Cross
Professor
Montana State University, Bozeman, MT

Eric Scholl
Postdoctoral Researcher
U.S. Geological Survey, Flagstaff, AZ

CONTENTS

CONTRIBUTING INTERVIEWEES

Caleb Bollman is a fisheries management biologist with Montana Fish, Wildlife & Parks located in Miles City. He has held this position since 2010 working on a range of fisheries from put-and-take recreational pond fisheries to Pallid Sturgeon recovery work on the lower Yellowstone River and its tributaries (i.e., Tongue and Powder Rivers). He holds a bachelor's degree from Milligan University, and a master's degree from Ball State University where he conducted his graduate research investigating Yellow Perch early life history on Lake Michigan.

David Brooks most recently served Montana Trout Unlimited as Conservation Director and became executive director in May 2017. He is a trained environmental historian who studied Superfund sites and watershed issues. As part of earning his Ph.D. in history at the University of Montana, David wrote the book on Superfund cleanup of the Milltown Dam site on the Clark Fork. Since moving to Missoula in 2000, he has walked, jogged, pedaled, paddled, rowed, hunted, skied, and ogled as many wild places and open spaces in the state as possible, most enjoyably when he's in the company of his wife, daughter, dog, and friends.

Matt Boyer works with Montana Fish, Wildlife & Parks in Kalispell. Some of his most enriching experiences gained from the fisheries profession are the friendships made at Montana Chapter AFS meetings and while afield in wild places like the Bob Marshall. His wife, Gretchen, and son, Reed, are his favorite partners in the outdoors. He received his bachelor's degree in wildlife and fisheries science from Penn

State University and his master's degree in organismal biology and ecology from the University of Montana.

Chris Clancy worked for Montana Fish, Wildlife & Parks from 1978 to 2019. He first worked for 2 years out of Miles City and Glendive and then as the management biologist for the upper Yellowstone River out of Livingston for 9 years. In 1989, he moved to Hamilton as a fisheries biologist on the Bitterroot River, retiring in 2019. He holds both bachelor's and master's degrees from Montana State University.

Pat Clancey worked for Montana Fish, Wildlife & Parks for over 34 years, retiring at the end of 2015. His first summer with FWP was as a game range laborer, then as a fisheries technician and biologist. He spent about 5 years in Kalispell working in the Swan Drainage and Flathead River and Lake, then 5 years at Fort Peck working on Pallid Sturgeon in the Missouri and Yellowstone Rivers and tailrace fisheries issues. He then worked in Ennis conducting native fish conservation & restoration projects and hydropower mitigation. He holds bachelor & master's degrees in fish & wildlife management from Montana State University.

Jim Dunnigan began his career as a fisheries biologist with the Yakama Indian Nation where he worked on several Pacific salmon restoration projects from 1996 to 2001. He grew a deep appreciation and respect for the special connection the indigenous people shared with the fish he worked to restore. In 2002, Jim moved to Libby after accepting a fisheries biologist job with Montana Fish, Wildlife & Parks on the Libby Dam Mitigation Project and served in that position for 19 years. Jim was promoted to Libby Dam mitigation coordinator in 2021 and currently remains in that position. Jim holds both bachelor's and master's degrees from the University of Idaho.

Wade Fredenberg developed a passion for fish while growing up in Kalispell, before he even knew what a fisheries biologist was. By chance, he met Bob Domrose – an early Montana Fish, Wildlife & Parks biologist, mentor and neighbor – and the die was cast. After two years at Flathead Valley Community College, he went on to a bachelor's in fish and wildlife management at Montana State University. Under the wing of Dr. Bill Gould, he also got his master's at MSU by 1980, launching him on a rewarding 40+ year career. Wade worked for Fish, Wildlife & Parks as a fisheries management biologist on the Musselshell and Bighorn Rivers out of Billings and then followed Dick Vincent as the Madison and Gallatin biologist in Bozeman. In 1992, he moved back to his home waters in the Flathead where he spent 25 years with the U.S. Fish and Wildlife Service, working on fish production evaluations at Creston Hatchery and then acting as the first USFWS Bull Trout coordinator. He considers Bull Trout to be the crown jewel in the Crown of the Continent.

Beth Gardner studied at Michigan State University. She began working for the U.S. Forest Service in northern Michigan and moved to the Flathead Valley in 1994. She currently serves as a fish biologist on the Flathead National Forest where she has worked to restore fish passage, protect native fishes from invasive species, and conserve stream habitat.

Bob Gresswell first worked in Yellowstone National Park in 1969, and the experience changed his life forever. Following his first hitch in graduate school at Utah State University, he returned to Yellowstone for 17 years where he was employed as a fisheries biologist and assistant project leader with the U.S. Fish and Wildlife Service. In 1990, Bob returned to graduate school at Oregon State University to work on a Ph.D. focused on the Cutthroat Trout of Yellowstone Lake. He started working for the U.S. Geological Survey in 1997 in Corvallis, Oregon and became a Courtesy Faculty member at Oregon State University. In 2004, he returned to God's Country, where he worked as a research biologist with the USGS Northern Rocky Mountain Science Center in Bozeman and served as an affiliate assistant professor in the Department of Ecology at Montana State University. He retired from the USGS in 2015 and is currently an instructor at MSU. Bob remains active in the American Fisheries Society and the Wild Trout Symposium series and is the chair of the science panel reviewing Yellowstone Cutthroat Trout recovery in Yellowstone Lake.

Barry Hansen has worked for the Confederated Salish and Kootenai Tribes since 1990 and with Montana Fish, Wildlife & Parks and U.S. Forest Service in Libby during the ten previous years. He received a bachelor's degree from Tulane University and a master's degree from the University of Montana. Preservation of Bull Trout has been a consistent theme throughout his working years.

Travis Horton began his career working for Idaho Fish and Game as a fisheries technician in 1993. In 2000, Travis worked for the University of Idaho researching salmon and steelhead migrations throughout the Columbia River Basin. Travis began working for Montana Fish, Wildlife & Parks in 2001. Travis started as the Smith and Missouri River biologist in 2001, transferred to the statewide native species coordinator in 2006, and finally became the Region 3 fisheries manager in 2010. In 2022, Travis became the director of environmental health for Gallatin County. Travis holds a bachelor's degree from the University of Idaho and a master's degree from Kansas State University.

Chris Hunter received a bachelor's degree in biological science from the University of California and a master's degree in zoology (limnology) from the University of Montana (Go Griz). He held a variety of jobs in Montana related to water quality, water rights and fishery issues for government agencies and as a consultant. He was hired by Montana Fish, Wildlife & Parks in 1989 to work on instream flow reservations. In 1991, Island Press published *Better Trout Habitat* which he authored with Tom Palmer. At that time, he was president of Montana AFS, and organized a national meeting titled "Practical Approaches to Riparian Management" with Bob Gresswell. He worked at MFWP for 20 years, the last 8 as chief. These days I travel with my wife, hike with geezers, build boats, and follow our 4 granddaughters around.

Mike Jakober spent the bulk of his career working for the U.S. Forest Service on the Bitterroot National Forest in western Montana. After completing his master's degree on Bitterroot River tributaries under the supervision of Dr. Tom McMahon, he became a full-time biologist in 1995. He worked closely with Montana Fish, Wildlife & Parks to collect data and complete fisheries projects until his retirement in 2023.

Carter Kruse is the director of conservation and science for Turner Enterprises, Inc. He works to conserve the species and landscapes across the Turner properties in the western and central U.S. He also works closely with partners from numerous agencies, organizations, and universities to conduct research and restoration projects. He holds a Ph.D. from the University of Wyoming.

Brian Marotz worked for Montana Fish, Wildlife & Parks for over 34 years, serving as both a special projects manager and hydropower mitigation coordinator in northwestern Montana. He spent much of his career focusing on mitigation of water-management impacts to native fishes. He holds a bachelor's degree from the University of Wisconsin-Stevens Point and master's degree from Louisiana State University.

Tom McMahon is a professor emeritus of fisheries at Montana State University. He taught and conducted research at MSU for 31 years, from 1990 to 2021, on a wide variety of applied research issues. He mentored 31 graduate students that now work throughout Montana and the U.S. He taught courses in fisheries ecology and management, ichthyology, introduction to fish and wildlife, fish habitat management, and research methods. He is grateful to have been able to stand on the shoulders of past forward-thinking Montana fisheries biologists, and to have worked with a highly competent, dedicated, and fun group of Montana fisheries professionals and students on a diverse set of interesting and challenging research questions.

Leslie Nyce grew up in Pennsylvania hunting and fishing with her father which instilled a love of the outdoors. Leslie worked in conservation education for many years prior to starting her fisheries technician job with Montana Fish, Wildlife, & Parks in the Bitterroot in 2002. She holds a bachelor's degree from Kutztown University, Pennsylvania and a master's degree in fish and wildlife from the University of Montana where her thesis work was on Bull Trout genetics in the East Fork Bitterroot River. She considers Bull Trout to be the crown jewel of the trout species.

Mike Ruggles, a native Montanan raised on a North Dakota farm, earned a bachelor's degree from Dickinson State College and a master's of natural resources from the University of Wisconsin-Stevens Point. Mike has worked with Montana Fish, Wildlife & Parks since 1994 on fisheries in the Missouri and Yellowstone River basins from Pallid Sturgeon and Yellowstone Cutthroat Trout, Walleye and Rainbow Trout, and a great mix of Sauger, catfish, and many other Montana fish. In 2021, he became the Region 5 supervisor which includes coordination with fisheries, wildlife, parks, communications and education, and administrative regional programs. He enjoys the diversity of fish, wildlife, and parks, and working with people on the landscapes in his responsibility areas and beyond.

Pat Saffel grew up hunting and fishing in South Dakota. He received a bachelor's degree from South Dakota State University and a master's degree from the University of Idaho. He has held his current position as the Montana Fish, Wildlife & Parks fisheries manager for Region 2-Missoula since 2001. Before that, he was a fisheries biologist in Thompson Falls.

Brad Shepard was born in Dayton, Ohio and headed west after high school. He received a bachelor's degree from Montana State University in 1975 and a master's degree from the University of Idaho in 1981. He later returned to MSU for a Ph.D. on Cutthroat – Brook Trout interactions, receiving the degree in 2010. Throughout his 40+ year career, Brad worked to conserve Montana's native fishes as a biologist, manager, researcher, and professor. Brad passed away in 2021, but his tireless work on behalf of native fishes (especially Cutthroat Trout), his mentorship of numerous grad students & technicians, and his service to Montana AFS will long be remembered.

Amber Steed began her career in fisheries as an intern in Yellowstone National Park, driving her to better understand and conserve aquatic species across the West. She has been a fisheries biologist for Montana Fish, Wildlife & Parks since 2007, focusing on the Flathead drainage in northwest Montana. She uses research, monitoring, and on-the-ground conservation projects to benefit native species while building community awareness and support for her work. She earned a bachelor's degree from the University of Wisconsin and a master's degree from Montana State University.

Jim Vashro worked for Montana Fish, Wildlife & Parks for 39 years. He was hired as a creel clerk, then a hatchery worker, then the fisheries management biologist for the upper Clark Fork for 6 years. He was promoted to the regional fisheries manager for Region 1 in northwest Montana, a position he held for over 30 years until retirement in 2013. He developed the wilderness fishing regulations for the Bob Marshall and Great Bear, developed and standardized conservative regulations for Bull and Westslope Cutthroat Trout, and created trophy trout lake regulations and family fishing ponds while focusing on aquatic education, fishing access, and illegal fish introductions. He earned a bachelor's degree from the University of Montana and a master's degree from Cornell University.

EDITOR, FOREWORD, AND ARTIST

Niall Gallagher Clancy grew up in the Bitterroot Valley of western Montana and has a strong interest in the conservation of native fish communities. He holds a bachelor's degree from Montana State University, a master's degree from Utah State University, and is currently working on a doctorate at the University of Wyoming. He currently serves as the co-chair of the Montana AFS Species of Special Concern Committee.

Ernest Richard (Dick) Vincent's first jobs with Montana, Fish, Wildlife and Parks was as a summer fisheries worker in Glasgow, MT (1960) and Kalispell (1961 and '62). In 1966, he was hired as a fisheries management biologist for development of techniques to sample and estimate fish populations in larger streams and rivers plus management of the Madison and Gallatin River drainages. From 1988-1996, he was promoted to regional fisheries manager for Region 3 out of Bozeman. In 1996, he was assigned by the department director to oversee whirling disease research in Montana. He retired in January 2008. He received both a bachelor's and master's degree in fish and wildlife management from Montana State University in 1963 and 1966, respectively.

David Ritter is a fish biologist for the U.S. Geological Survey researching Pallid Sturgeon. He is passionate about fisheries science and technology and the conservation of aquatic resources and native fishes. He spends much of his spare time creating fish and wildlife art (www.rittercraft.com). He holds a bachelor's degree from Virginia Tech and a master's degree from Montana State University.

FOREWORD

When I look back on my decision to become a fisheries biologist, I really had no idea what a fisheries biologist's duties were. I associated the job with working outdoors and with fish. In order to learn more about this career choice, I contacted the local game warden, who gave me some ideas of what a fisheries biologist's duties were and what type of education was necessary. He did recommend Montana State University (College) at Bozeman, Montana, which offered B.S. and M.S. degrees in fish and wildlife management. This turned out to be one of the best decisions that I ever made.

When I graduated from MSU in 1966, I was ready to be a fisheries biologist. I was offered a newly formed position to develop techniques, which included creating sampling equipment and a system to estimate the fish abundance within certain reaches of the stream. Up to this point in time, current fish population estimates were limited to small streams where they were captured via electrofishing techniques and counting what was captured. I felt that I was adequately prepared for this research project from what I had learned at MSU. But, there was more to this position than research. It included managing the fisheries for the Madison and Gallatin River drainages including dealing with habitat issues, water flows, angling regulations, angling clubs, local businesses invested in deriving income from the fisheries, and any other group or agency with interest in the fishery resource. This turned out to be an area for which I was much less prepared. Here, I had to rely on more experienced fisheries professionals for advice and direction. This type of interaction is important for new fishery professionals, as well as the more experienced. I recognized that there was a need to improve areas where I was weak, such as public speaking and dealing with sportsman groups and public agencies, in order to be more effective.

I relied on wildlife professionals that were successful in these areas, such as Art Whitney, John Gaffney, LeRoy Ellig and Ron Marcoux. Art Whitney (chief of fisheries) taught me to always stick to your principles, and be willing to stand by your data even though it may be unpopular at the time. An example was the research study to determine the impacts of stocking hatchery Rainbow Trout on wild trout populations, the results of which suggested stocking was detrimental to wild trout populations. This study was very unpopular with some angling groups, some local businesses relying on angling, and some groups within the Fish, Wildlife and Parks Department - it became very political. Art supported this study based on its scientific value and could have denied the project as too political. The same was true for John Gaffney (regional fish manager) and LeRoy Ellig (regional supervisor). I admired these qualities and tried always – To Follow the Data!! I always admired the skill that Ron Marcoux exhibited in dealing with the public and public agencies. Ron excelled in this area, and I learned a lot from him. You can always benefit from observing and absorbing the talents of your co-workers. The most difficult part of my career was fisheries' politics and choosing to retire. But would I choose this career again if I had to do it all over? Absolutely. It was one of thc best things that ever happened to me, besides meeting and marrying my wife, Twyla.

Dick Vincent
Montana Fish, Wildlife & Parks - Bozeman, retired

PREFACE

In 2019, I had the opportunity to sit in the press box at a minor league baseball game in Salt Lake. Many players on the field had already played in the majors and were either trying to gain a few more skills before permanently going to "the show" or to eke out a few more years as a professional ballplayer before calling it quits. Most of the players had spent several years in the "farm system" – a farm system consists of all the minor-league teams that a major-league team such as the New York Yankees or Boston Red Sox operates to train players and find the best new prospects.

In many ways, colleges and graduate schools serve as the farm system for the fisheries profession. Young outdoorspeople take a few classes their freshman or sophomore years and, if they can pass the wicked curve of organic chemistry and genetics, they'll likely end up with a degree in ecology or some related field. With a sufficient GPA, a few summers of experience, good references, and perhaps the addition of a well-written thesis, you have a decent shot at getting drafted by an agency or non-profit. If your numbers aren't very good, you might have to settle for a position as a starting pitcher with the Baltimore Orioles.

I was lucky to grow up in a household where fisheries and science (and baseball) were regular topics of conversation and to join my dad, Chris – a now-retired Montana Fish, Wildlife & Parks fisheries biologist (see Ch. 2) – as he, Leslie Nyce (see Ch. 7), Mike Jakober (see Ch. 3), and others managed the fishes of the Bitterroot. The knowledge I absorbed over twenty-or-so years has been invaluable as I myself work in fisheries science. These aren't technical skills, per say. They are observations, anecdotes, perspectives, and philosophies gained over a forty-year career. After attending my first few meetings of the Montana Chapter of the American Fisheries Society, I came to appreciate that every biologist and technician in the field has their own unique set of experiences and perspectives, but this information was only occasionally passed on to the next generation of fisheries professionals – and even then, usually over beers. This publication is an attempt to record and summarize some of those perspectives so that they can be provided to a larger number of us early career professionals.

Beginning in 2018, I asked twenty-three individuals with broad experience in fisheries to provide answers to my questions about the perspectives and advice they'd like to share. They are the primary authors of this book. While I mostly collected interviews from the most experienced and longest-serving, fisheries-management professionals in the state (with a couple researchers sprinkled in), I also included several younger professionals to broaden the perspective, especially those from federal or non-profit groups. Recognizing that fisheries biology is also extremely male-dominated, I further endeavored to interview several women under 50 to provide a more balanced view. Many other talented professionals were not interviewed due to time constraints (theirs or mine), significant overlap with other interviewees, or my lack of familiarity with their work circa 2018. Their perspectives should surely be sought from those working around them.

I am very grateful to all contributors and have learned much in the process of editing this work. I am certain others will as well. The *Last Best* State and its fishes are lucky to have drafted such an all-star team.

Niall G. Clancy

Arctic Grayling (*Thymallus arcticus*)

CHAPTER 1
Values and Skills

Interviews with

Bob Gresswell, Amber Steed, Brad Shepard, Matt Boyer, and Beth Gardner

The fisheries profession demands we learn how to gather, analyze, and disseminate data. But all of these skills are secondary to the most important attributes of great biologists: motivation and a deep respect for the people, places, and organisms placed in our care.

I asked contributors to this section to address the set of values and skills needed to be an effective biologist. Bob Gresswell and Amber Steed were asked to provide their motivations for working in fisheries and the philosophies that help them address important issues. Brad Shepard, Matt Boyer, and Beth Gardner were asked to discuss what skills, technical and especially "soft" skills, they believe are most needed in the profession.

Bob Gresswell

Montana State University - Bozeman
U.S. Geological Survey, emeritus
MTAFS Chapter President 1988
MTAFS 1989 Outstanding Professional Award

In June of 1969, I drove north from my home in New Mexico to Yellowstone National Park to work for the U.S. Fish & Wildlife Service. After a summer in the park, the Yellowstone Cutthroat Trout had become part of my soul. My early experience there gave me a strong sense of wonder about functioning natural ecosystems. Those formative years in my career solidified my view that we humans must conserve all of our remaining, properly-functioning aquatic ecosystems.

I subscribe to a very contextualist worldview that assumes the natural world is strongly enmeshed in a complex web of factors that vary in space and time. Because of this complexity and variability, we humans must be aware that we can never know absolute truth, and the best we can expect is a probabilistic understanding of the natural world. I believe that because of our imperfect understanding, we should be cautious with resource extraction (use) and conscious of potential unintended consequences of our actions. By definition, all biota influence the environment that they live in, but we humans should attempt to live in harmony with natural systems, not control them. Sustainability should be a guiding principle in cases where resource extraction is necessary, and preserving some areas from consumptive uses is critical not only for the survival of the biosphere, but necessary for the human soul. As Aldo Leopold suggested, the first rule of conservation should be "do no harm."

As I have gained knowledge, I believe that I have become more aware of the complexity of living systems, their variation in space and time, and their capacity for persistence. I recognize that resource extraction is necessary and can potentially be sustainable; however, human hubris and insatiability are attributes that often hinder attainment of that goal. I have a very strong protectionist view for fisheries and believe that preserving remaining intact systems should be one of our primary management goals. Undoubtedly, some systems cannot be restored, but a high priority should be given to systems that can be renovated. It is also important that we recognize that many of these altered systems can be a reservoir of potential invaders into unaltered portions of larger systems.

Although I often refer to functioning natural ecosystems, I recognize that humans are inextricably part of these systems. Therefore, it is critical to work with the public to engender a stronger appreciation of the importance of protecting and conserving our natural resources. Of course, this is easy to say but hard to achieve. It starts with education about, and appreciation for natural systems, and this should be embedded in the roots of our educational system. Knowledge about living systems is essential for their management. Furthermore, since managers and conservationists are involved with maintaining our natural resources, we must recognize that this includes resource extraction (use) and including the public in the formulation of management strategies is critical to the success of these strategies. Where consensus can be reached among disparate user groups, the probability of success increases.

--

Amber Steed

Montana Fish, Wildlife & Parks - Kalispell
MTAFS Chapter President 2018

As a kid growing up in Minnesota, I was happiest when exploring the woods and waters around my suburban home. Fishing at dawn with my dad for sunnies, perch, and bluegill had me fixated for hours. Collecting and raising tadpoles, finding bird nests, investigating dead things – these experiences and many more fueled my fascination for the natural world.

My work as a fisheries biologist has afforded me a direct connection with the resources that I strive to protect and with those who care just as deeply for them. Creating and perpetuating a bond with nature, through fishing and other outdoor recreation, gives me hope that our future citizens and decision-makers will want to protect those experiences the way I do.

While not all of us are management biologists (myself included), having a management philosophy in mind is important to guide how we approach the challenges and opportunities of working in fisheries conservation. My approach when facing a new or dynamic project or decision-point is to ask the following questions: (1) Is what we're doing now going to make a substantive, positive impact on the future of these resources? (2) What course is most sustainable - socially, biologically, economically? How does the best available science inform that? (3) Could this action improve or maintain the connections (memories) that people make with aquatic resources and with one another through their experiences outdoors?

While many of the issues we face as fisheries biologists are complex and messy (e.g., nonnative fish removal or multi-jurisdictional coordination), I always try to distill issues down to the parts that matter most. It can be easy to fixate on the details of project feasibility – we don't have the funding for that, or this approach has never been done, etc. Stepping back to see the broader picture, to help focus on what matters most, guides me in the right direction. Perhaps most importantly, I try to never let the fact that something has always or never been done in a certain way dictate how I move forward. Where would we be now if that was our guiding principle?

Further, every project and decision can benefit from the input of others – my coworker across the office working in another drainage, our crew of technicians, managers, and those outside our work unit. Innovative ideas, fresh perspectives, and diverse experience tend to improve every project I have tackled, building trust and better working relationships in the process. Some great ideas have even come out of questions from kids during career fairs or presentations, reminding me that we should always remain open-minded and curious in how we approach our work.

We're often expected to be experts in everything – the biology of our subject species, their habitats, as well as communicating about what we learn and think people should care about. It can be scary to admit when you don't know all you think you should about something, or aren't as skilled at some part of your job as someone else (almost always the case!). Acknowledging when you are or aren't confident in something is not a weakness, but a strength.

--

Brad Shepard

Montana Fish, Wildlife & Parks, retired†
MTAFS Chapter President 1996
MTAFS 1990 Outstanding Professional Award

†Brad Shepard passed away in September 2021. A remembrance was published in the November 2021 issue of *Fisheries.*

In my opinion, the fisheries profession is not just a profession, it is a way of life. The first thing a person must have is the personal commitment to become the best professional fish scientist or worker they can become. The opportunities for employment in the fisheries profession are so limited, we must have the very best people in those positions. Are you ready to commit your life to this profession? Are you willing to "speak the truth to power", even if doing so puts you at personal risk? I can almost guarantee that if you go into the fisheries profession, you will face this type of choice sometime during your career. Will you step up?

Secondly, I don't want to ignore the very real need for a person to know about the science and how to do the job technically. Usually, this aspect is adequately covered in most college curricula, but a foundation in basic sciences with advanced education in population ecology, life history, taxonomy, evolutionary ecology, conservation ecology, population and people management, and genetics are all very important. Gain statistical knowledge, so you can analyze your data and talk to high-powered statisticians, as necessary.

Develop good communication skills, both verbal and writing. I think these skills are often under-emphasized in the formal education of fishery professionals. Communicating often difficult and complex scientific concepts effectively to your peers, your bosses, and the public is critical to doing a good job. Learn how to do this and practice, practice, practice. Take every opportunity you can to speak in public and write popular and scientific papers. Ask those folks whose articles you read and enjoy, or are impressed by, to review your writing. Same goes for folks who impress you with their oratory skills. Learn from those you want to emulate.

Learn how to work in team settings. Most management decisions are made in team settings and you need to know how to work and accomplish things in these types of settings. Learn how to be effective in organizations, as this will pay dividends in how you interact with your employer's organization but will also help with all interactions you will *have with formal and informal organizations during your* career. One of the best "continuing education" workshops my employer (Montana FWP) sponsored me to attend explained the many different ways individuals have for influencing decisions and wielding "power". I found out that a person did not necessarily have to become a boss (supervisory position) to wield a tremendous amount of power, rather they could develop a network of peers who shared a common vision, and commitment to achieving that vision, and all they needed to do was act on that shared vision.

Use your communication skills and team experiences to develop long-lasting personal and professional relationships that are built on mutual trust, respect, and shared values and goals. I have found that many of my very best friends are people I have worked with over the years. I believe that these personal and professional relationships allowed us to accomplish significant fish management and conservation actions in the past two decades. I seek out some folks I do not know at larger gatherings (i.e., AFS meetings) and just talk about fish, fishing, and life in general. I usually leave each gathering with one or two new colleagues. Young professionals and students often feel intimidated by older professionals, but I can assure you, older professionals will feel honored if you ask their opinion or ask them about their work. Find a professional who is willing to become a mentor to you and develop a formal mentor-mentee relationship. These relationships can be very valuable. When you feel like you are ready, become a mentor for other younger professionals.

Learn how to listen and critically review the scientific literature. Preparation for a profession like fisheries is like any other type of preparation – you can learn about it from the experience of others, you can be trained to do it by those who know how to do it, or you can experience it yourself and learn from your experiences. I submit that almost all of us have used all three methods above, but the more we learn and train, the better off we are to handle any given situation and the fewer mistakes we are likely to make when we encounter a situation for the first time. Many of the questions you are asking, or are being asked of you by your employer, have been investigated by someone in the past. Learn how to find and critically review both scientific and popular information and create a habit of doing this on a regular basis. Become an expert on a few topics, but maintain your general knowledge, and never stop learning. Put your personal observations in context with observations by others, but also in geographic and temporal context (location, time, and scale likely matter).

Take advantage of opportunities to present your information whenever and wherever you can. Formally presenting your information helps you build your communication skills, but also is an obligation of anyone doing science. If you do not share your information, why are you wasting resources collecting information? Every field-oriented professional scientist should publish at least one peer-reviewed paper every five years, even those in management positions. In my opinion, it is part of the obligation we all have for getting to have so much fun collecting data. If you collect the data, analyze and summarize it every year. Keep up with your data. If you can't keep up with it, don't collect it.

Take advantage of available opportunities to continue your education (keep learning) and don't be afraid to attend education opportunities that may not seem pertinent. When evaluating educational opportunities, be discerning, but don't be afraid.

--

Matt Boyer

Montana Fish, Wildlife & Parks – Kalispell
MTAFS 2018 Outstanding Professional Award

By the time this collection of interviews is published by the Montana Chapter of the American Fisheries Society, I will have worked in the fisheries profession for more than 20 years, first as a field technician, then as a project biologist, and currently as program coordinator for hydropower mitigation in northwest Montana. Those years feel to me as though they have gone by quickly, and I am extremely fortunate for the opportunities that have come my way and, especially, for the chance to work with so many incredibly talented and passionate people.

I didn't grasp it at the time, but thinking back on my undergraduate and graduate programs, it's now apparent how every one of the courses I took in those diverse curricula provided something to help prepare me for a job in fisheries.

While in school, it was obvious to me how the material in classes like aquatic ecology and limnology was going to be directly relevant to a fish biologist's daily job duties. Courses like calculus, physics, genetics, and organic chemistry seemed somewhat less pertinent; nevertheless, I understood them at the time to be essential for developing a solid foundation in the 'hard sciences'. Working toward a diverse scientific technical understanding should be a top priority when selecting courses from your program's curriculum. The additional academic challenges you'll face in those courses also help promote good work habits for the future.

Use your electives wisely! Sure, it's tempting to fill a couple credits with a golf or tennis offering, but why not take advantage of an independent study working with a Ph.D. student on their grad project or apply for that in-between-semester internship with an agency or non-profit? These opportunities provide hands on experience, help develop relationships and grow professional networks, and they might even lead to future employment prospects.

What about courses like social psychology, renewable resource economics, effective communication, environmental philosophy and ethics? I enjoyed these 'soft science' courses but, at the time, saw them as little more than necessary credits to obtain the degree. In my naïve view, they seemed tangential to what I envisioned fish biologists would be doing in their day-to-day work. After some time on the job, I realized their critical role in developing well-rounded fisheries professionals possessing what are known as 'soft skills' – a set of personal attributes and abilities that allow you to effectively interact with others in the workplace. For many, soft skills are most effectively honed through on-the-job life experiences. However, academia plays a vital role in helping future fisheries professionals succeed in this area. Courses and seminars in the field of conservation biology, for example, highlight case studies where scientific principles and theory blend with social, political, and economic dimensions to achieve aquatic resource conservation. A fisheries professional that can skillfully interact with their peers and the public will bring more project ideas to fruition and enjoy greater job satisfaction.

There are more options than ever to engage in activities that distract us from nature and the outdoors. For many, this translates to more time in front of a screen and fewer experiences of the type that develop genuine curiosity for and understanding of the natural world. I would like to see our fisheries profession grow more naturalists (think Charles Darwin or E.O. Wilson), even if that means we have fewer experts in Bayesian stock assessment (though we absolutely need those, too!). As professionals in our field become more specialized in their area-of-expertise, we run the risk of not cultivating an innate ability to think in a way that crosscuts multiple disciplines to develop conservation approaches with broad influence and lasting effect. Take school and your job seriously, but be sure to make time to enjoy the natural resources you're helping conserve!

--

Beth Gardner

U.S. Forest Service - Bigfork

The best skill sets are a (1) solid understanding of fisheries ecology (population dynamics, habitat, etc.), (2) verbal and written communication, (3) problem solving, and (4) general natural resource management. I think a person with these four key skills will thrive in any setting. They will be valued by any employer in any location.

Fisheries ecology is probably the most readily grasped. This includes things such as population estimation, habitat restoration, limnology, etc. This is the technical stuff that catches our interest in the first place. Solid technical skills can be obvious to the public and generates respect and trust. Communication is harder to define, but it includes the ability to explain science to the public (and managers), the ability to collaborate and persuade, and also being responsive. Problem solving means finding creative solutions to resource issues, mechanical breakdowns, and logistic headaches that always go with fieldwork. Some people might call problem solving, "leadership", but I think there is a difference. Leadership is the ability to inspire and direct human resources, and that is awesome, but not every biologist needs to be a leader. General resource management is having a good understanding of hydrology, geology, wetlands, forestry, etc., and this greatly helps in finding solutions to resource problems.

Other than perhaps problem solving, all of these skills are obtained through both formal training and experience. Technical skills in fisheries and natural resource management, of course, require formal training, but they are greatly enhanced by actual field experience. No matter how great the classroom education, it is the actual experience that refines and improves the skill set. Communication needs some formal training, but most of it is learned by mentoring and experience. Problem solving is the exception. This cannot be formally taught but is gained by experience, be it fisheries work or any other supervisory experience.

Young biologists generally have a good understanding of natural resource management, good fisheries ecology skills, and some are proficient in GIS or writing, but most lack verbal communication skills and problem solving. They struggle to explain the science to the public and flounder when a logistic challenge rears its head. That is not necessarily a great failing, just part of the learning experience. A young biologist needs a good mentor to help them thrive.

The best advice I ever got was to get lots of field experience. Having good grades is good, but having field experience is vital. While I was an undergraduate, I was urged to accept seasonal work no matter how low paying and short-term. So, I did. I worked my summers at various temporary positions with the National Park Service, university, and U.S. Forest Service. When I finished my bachelor's degree, I had quite a suite of experience. I could wrangle gill nets, repair electrofishers, inventory trout streams, classify wetlands, identify fish species, and more. It showed that I really got into this kind of work, and I was flexible enough for a new challenge. I still had much to learn when I finally landed a permanent biologist position, but at least I was on my way.

--

CHAPTER 2

Building Relationships to Conserve Cutthroat and Other Species

Interviews with

Pat Clancey and Chris Clancy

Cutthroat Trout conservation and restoration has been one of Montana's primary fisheries-interest projects since the era of native species conservation began in the 1970's. While the procedure of chemically removing invasive species with the piscicide rotenone and restocking with native trout is now down to a fine art, the following two examples, written by the elder Clanc(e)y biologists[1]*, are unique – the Cherry Creek project because of the precedent it set for the state, and the Overwhich Creek project because of its break from that precedent. I asked both to explain the impetus for the projects and provide some of the philosophical underpinnings and social challenges experienced along the way.*

---

[1] Pat and Chris use different spellings of Clanc(e)y, but they are in fact brothers. The editor's uncle and dad, respectively.

Pat Clancey

Montana Fish, Wildlife & Parks, retired
MTAFS Chapter President 2002
MTAFS 2010 Outstanding Professional Award

An extensive and detailed presentation of the Cherry Creek Project can be found at:

Clancey, P.T, B.B. Shepard, C.G. Kruse, S.A. Barndt, L. Nelson, B.C. Roberts and R.B. Turner. 2019. Collaboration, commitment and adaptive learning enable eradication of nonnative trout and establishment of native Westslope Cutthroat Trout into one-hundred kilometers of Cherry Creek, a tributary to the Madison River, Montana. Pages 589-647 in D.C. Dauwalter, T.W. Birdsong, and G.P, Garrett, editors. Multispecies and watershed approaches to freshwater fish conservation. American Fisheries Society, Symposium 91, Bethesda, Maryland.

Montana FWP had been documenting the status of native Westslope Cutthroat Trout (WCT) in the Missouri River Drainage, east of the continental divide, since the 1980's. It became clear that they had been extirpated from all major rivers within the Missouri Drainage and were generally restricted to headwater areas of tributary streams where streamflows generally are less than 5 cfs. Except where there was a natural or man-made barrier isolating the WCT population, they were at risk of invasion by non-native trout such as Rainbow, Brook and Yellowstone Cutthroat.

In 1997, FWP crews began surveying tributaries of the Madison River with the intent of to identify streams that would be suitable for WCT re-establishment. Ted Turner, founder of CNN and owner of the Flying D Ranch near Bozeman, read an article about our work in the newspaper and set up a meeting between his organization, Turner Enterprises, Incorporated (TEI), Dr. Calvin Kaya at Montana State

University, and FWP. After discussion, FWP began surveys of the Cherry Creek Drainage in late August, 1997. A key feature of the Cherry Creek Drainage is that a 30-foot waterfall exists in Cherry Creek Canyon, about 8 stream miles above the confluence with the Madison River. Above this waterfall we identified approximately 60 miles of streams, mostly tributaries to Cherry Creek, that were occupied by trout. Approximately 15 miles of stream were unoccupied by trout due to natural barriers that prevented invasion. The only fish species we found during surveys were Rainbow Trout, Brook Trout and Yellowstone Cutthroat Trout. Due significantly to the high-quality aquatic habitat, total length of stream above the waterfall, and the pledged support of TEI, we concluded that Cherry Creek was an opportunity too good to pass up. We began drawing up plans to remove the non-native fish and replace them with native Westslope Cutthroat.

In the late 1990's, native species conservation was taking hold in the fisheries profession across Montana, in the Montana Chapter of AFS, in federal agencies, and in some angling organizations. FWP's native fish program had been established but had not yet had significant impact on the ground. However, the motivation for the program was strong, relying not only on an "it's the right thing to do" philosophy, but also on legal grounds that if the WCT or other native fish species were listed as a threatened or endangered species, Montana may lose the authority to manage those species to the federal government. FWP's mandate is to manage all fish species within the state, so ignoring the status of imperiled species is contrary to that mandate. WCT had already been petitioned at least once for listing as a threatened species, and at least two more attempts would be made in coming years.

An interesting philosophical situation we were faced with regarded the application of the piscicides. I don't know a single one of the principal biologists who thought putting any chemical in streams or lakes was acceptable, and we were faced with the conflict of intentionally putting a chemical in the stream that would kill fish. We developed an empathy for people who work in industries that pollute, but I suppose we felt our cause more noble than just making money. We understood the larger picture of our intentions, and that the piscicides would be very short-lived, persisting for a matter of hours, not chronically on-going like so many industrial chemical applications.

The delay, 1998-2003

Pushback to the project delayed its implementation for several years and was generated primarily from a handful of individuals who had a history with the Flying D Ranch prior to Turner's ownership of it. At one time, the ranch apparently was pretty open to access for hunters and anglers, but eventually an owner previous to Turner closed that access. This handful of sportsmen, who did many good things for hunters and anglers through their organizations, went to battle with the Flying D over that closure of access. When Turner purchased the ranch, his name elevated any and all issues associated with the ranch. Local and regional media began covering stories related to the access battle waged by the sportsmen, and when the Cherry Creek Project was proposed, the issue immediately became controversial for several reasons: removal of game fish, use of piscicides in an outstanding water, FWP cooperating with a rich, out-of-state landowner who did not allow public access, and entering into a financial arrangement with that landowner to support an FWP biologist for the Cherry Creek Project.

Local media closely covered the story, and the case can be made that they actually fanned the flames of controversy to some degree due to the tone of some of their articles. Numerous letters to the editor on both sides of the issue were published. Several guest editorials supporting the project were published, but the primary project opponents made false claims about many aspects: that the piscicides would cause deformed babies to be born to women in Three Forks, that the project area would be left devoid of life by the piscicides, that FWP biologists were sidling up to the Turner money-trough and selling out a public resource (the non-native trout in Cherry Creek) to get rich, that the project was part of a plan for a United Nations take-over of the U.S., and so on. National and regional publications and magazines also covered the project, some objectively, some not.

Ultimately, as things moved forward towards implementing the project, opponents began complaining to politicians– including the Three Forks mayor and city commissioners, Gallatin County commissioners (though the project area was entirely within Madison County), three different Montana governors, all three Montana Congressional representatives – and trying to generate negative public opinion about the project. I appeared before numerous sportsman's & civic clubs and groups to present and discuss the project, as well as appearing before the Three Forks mayor and city commission, on local FOX talk radio in Bozeman (I was on a live one-hour call-in show that aired immediately before the 3-hour Rush Limbaugh Show), on the Face the State television show, as well as interviews by most print media or college classes that were doing a story or studying the project. Complaints to the Governor's Office were routed to FWP, eventually to me, for a response under the Governor's signature. None of the three Governors in office over the duration of the project opposed it, and Marc Racicot even vocally supported it. All three of Montana's members of Congress directed their staff to inquire about the project but determined that the federal government had no authority to terminate the project and that the Gallatin

National Forest had properly conducted their internal processes, and I think at least Senators Baucus and Burns favored the project.

One state legislator from the Bozeman area planned to introduce a bill into the legislature that would specifically identify and prohibit expenditure of state money on the Cherry Creek project. FWP administrators feared the bill would then be expanded by other legislators to prohibit many other actions to conserve or restore native fish. Then FWP Director Jeff Hagener (another Havre native) made a deal with the Bozeman legislator that FWP would not spend state money on the project if he would not introduce his bill. That left TEI to carry even more of the financial cost of the project, which they did willingly. Despite all the negative chatter and fear of the project, most people who heard objective presentations and discussions about it came away understanding the reason for the project, thinking it was not a big deal, and became ambivalent or supportive.

As the political and public opinion routes failed to derail the project, the opponents went after the various authorizations and permits we needed for applying the piscicide to the streams and lake. Their first step was to file appeals to the Gallatin National Forest and Montana Department of Environmental Quality (DEQ) against their authorizations of the project. Despite a statute in Montana law that allows for nuisance aquatic organisms to be controlled with aquatic pesticides, DEQ suspended our permit so they could do an Environmental Assessment of the project. DEQ held a public meeting in Bozeman in July 1999, and ultimately issued their EA approving and re-issuing the permit in October. Within 15 minutes, the attorney for the project opponents appealed the DEQ decision to the Montana Board of Environmental Review (BER; DEQ's citizen commission). The BER appointed a hearing examiner to review the appeal, a process which took nine months. After the hearings examiner recommended that the BER deny the appeal in its entirety, the BER issued its denial but also issued a 33-day stay to allow the opponents the opportunity to file a lawsuit against the project. On the 33rd day of the stay, the opponents filed a lawsuit. Opponents made numerous claims in the suit similar to those contained in their DEQ appeal, including that DEQ was allowing FWP to violate state and federal clean water rules. Almost 1½ years later, the Montana Court ruled in favor of FWP, but opponents then re-filed their suit in federal court. The Federal District Court of Montana eventually ruled very strongly for FWP.

In 2003, we were finally allowed to begin implementation of the project, and a final decision by the Ninth Circuit Court of Appeals in 2005, ruling for FWP on all points, ended the many challenges to the project.

The commitment

Neither I nor any of the principal biologists from the Gallatin National Forest or TEI ever doubted this project was the right thing to do. Nor did any of the FWP legal staff that were involved, nor any of the FWP administrators in the Helena state headquarters. And, as far as I know, there were no decision makers within the Forest Service, either at the forest or the regional office in Missoula, or at TEI, including Mr. Turner, that reconsidered their support for the project. Within FWP, there were two administrators in the Bozeman regional headquarters that thought FWP should abandon the project. Both had been wary of it from the beginning due mostly, I think, to their desires to avoid anything controversial. However, each of these individuals did support and defend the project in public forums whenever necessary. As the project proceeded, participants and decision makers became more committed to the project. This commitment from all project partners was the key to the project's ultimate success.

The upshot, 2004-2014

The eradication of the non-native trout was completed in stages down the drainage from 2003 – 2010, and establishment of a genetically pure WCT population was completely successful. WCT introductions occurred at dispersed sites throughout the project area (see map) from 2006 through 2012 and in 2014, primarily through the use of remote site incubators (RSIs) to incubate eyed eggs from five wild donor sources from within the Missouri River Drainage, as well eyed eggs from the state's WCT brood at Washoe Park Hatchery. The Sun Ranch, located in the south end of the Madison Valley and owned at that time by Roger Lang, developed a small private hatchery and rearing pond to support the Cherry Creek Project. Spawning fish in the donor streams were captured and spawned on-site with the fertilized eggs transported to the Sun Ranch Hatchery. There they were incubated until eye-up, then transported to the RSIs on Cherry Creek. Some of the eggs from each donor source were hatched at the Sun Ranch and introduced into the rearing pond to allow for mixing of the wild donor populations. As the fish in the Sun Ranch Pond matured, eggs and sperm were collected from them and mixed without regard for their original stream, then reared to the eyed stage, or to fry, for introduction into Cherry Creek. WCT fry from the wild donor sources at the Sun Ranch Pond and triploid (sterile) fry from the state's Washoe Park Hatchery were introduced in some locations in the project area where they would not affect on-going research related to the WCT RSI introductions, or to jump start recreational WCT fisheries within the project area.

The fish sizes and density of the resulting WCT population meets or exceeds that of the non-native trout present prior to the project. Angler catch rates for WCT in portions of Cherry Creek are one fish per minute. WCT have populated Cherry Creek below the waterfall and are being caught by anglers in the Madison River as far as ~30 miles downstream from Cherry Creek. No non-native fish have been sampled or detected in the project area since the final year of piscicide application in 2010. No introductions of WCT occurred above the natural barriers on those stream segments that were fishless prior to the project, so they remain fishless.

I think several key elements that helped me throughout the entire Cherry Creek saga, but especially during the period prior to the actual project when all sorts of challenges and professional & personal criticisms were being levied at me and others proponents, are that I was able to remain calm, responded to criticisms factually and professionally without retribution, and generally maintained some level of respect for opponents by recognizing that my only impression of, or interaction with, many critics was through the project. I tried to keep in mind that if I met this person in a different circumstance, such as a next-door neighbor or while watching a ball game in a bar, I might find them to be a "great guy" and have a very different impression of them. I believe these behaviors reflected well on my personal & professional credibility and also served to discredit the criticisms of those who were shrill or whose claims were marginally believable, or worse.

--

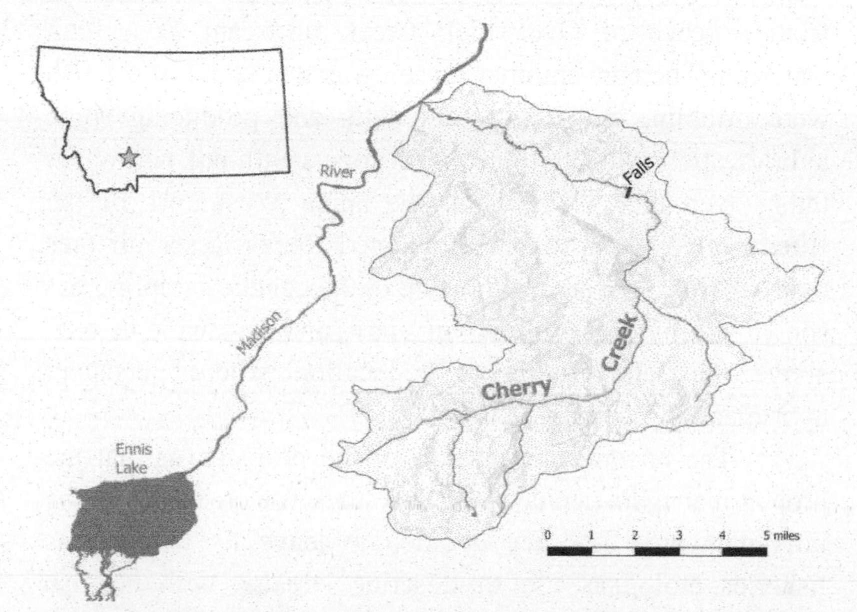

The location of Cherry Creek in the lower Madison River basin.

Chris Clancy

Montana Fish, Wildlife & Parks - Hamilton, retired
MTAFS Chapter President 1994
MTAFS 1993 Outstanding Professional Award

In many ways, the Overwhich Creek fish-removal project was routine. Our goal was to remove non-native Yellowstone Cutthroat x Westslope Cutthroat Trout hybrids from a reach of Overwhich Creek upstream of a high waterfall. The conventional reason was that some of the fish were dribbling down over the falls and producing some hybridization with the pure Westslope Cutthroat below the falls. Overwhich Creek is in the upper West Fork of the Bitterroot River, upstream of Painted Rocks Reservoir (see map). In this large area, almost all of the genetic sampling has shown that hybridization is very rare, and Overwhich Creek above the falls is the only location where persistent hybridization exists.

The unconventional part of the project was that the project managers decided not to restock the creek above the falls with any fish. We decided to leave it "barren", as fisheries biologists use to describe streams with no fish (barren sort of implies that, if there are no fish, it is of lower status). We felt that leaving this 10-plus miles of stream without fish was an ecological-restoration project. It is very unlikely that fish naturally inhabited this reach due to the large waterfall. Even with the non-native fish present, a decent number of tailed and spotted frogs were present. We saw other amphibians also. So, we felt that this reach should be inhabited by the species that were truly native, and that does not include fish.

The process of removing the fish using rotenone, while a significant amount of work over 3 seasons, was not particularly noteworthy. Our goal was to get all of them. If we were going to restock this reach with fish, we may have only treated it twice and then "swamped" it with nearest-neighbor Westslope Cutthroat Trout, of which there are many nearby sources. That may still happen if future monitoring reveals we were not successful in removing all of the fish and they begin to reproduce again.

So, why were we willing to work this hard to remove all of the fish to the point of spending an extra year on the project? What did the other folks who worked on the project think, and what did the public think?

Initially, there was some resistance from a few individuals and organizations. However, after we explained the reasoning, almost all of them agreed with the goals. I don't recall any agency people that expressed concern about leaving the stream fishless. One prominent fishing organization even suggested we treat the reach again if we found fish after three treatments.

A rhetorical question might be: Is a native species still native if it was introduced there by people? Would Westslope Cutthroat Trout be considered native when stocked in a mountain lake or above Overwhich Falls?

During my career, many things have changed in Montana, but two of the most obvious advancements were the cessation of stocking of catchable trout in Montana streams and the increasing profile and recognition of native species. The cessation of stocking occurred in a very hostile environment by then Montana Fish and Game in the 1960's and 1970's, just before I began work. The promotion of native species was more subtle and less controversial. From my perspective, agency biologists were a leading force in both efforts. I am hoping that the next generation of fisheries professionals brings a more "ecological" approach to fisheries management. I think Montana FWP is at a good place with its hatchery program. The hatchery program has excellent people, who are flexible and willing to do what needs to be done. The vast majority of the fish are stocked in lakes, reservoirs and ponds. Flowing streams are largely not stocked. That is a good balance.

The need for advancement is with the biologists. We value native species, but really only put much effort into the native sportfish. In many ways, I was just another trout biologist, of which there are many.

I worked almost my entire career with Montana Fish, Wildlife and Parks. I have great respect for almost all of the people I worked with. They are intelligent, committed and professional. I also worked with scientists from other agencies, universities and the private sector. We should realize that the organization we work for has a culture and the culture can advance. If you spend all of your time interacting just within your agency, it can be easy to fall into the "we know best" trap. The professionals in other agencies and the private sector are generally just as well educated, committed, and as intelligent as you are. Working with people in other agencies or organizations helps to diversify your views. If you go to AFS meetings, I suggest joining a committee and getting to know some people with different experiences. Younger professionals should broaden the gamefish culture of their agency. We took ecology classes for a reason. The focus of efforts will continue to be gamefish, but with little extra work, agencies can collect data on the fish that may not be pretty, grow large, take a fly, or taste good.

In my opinion the biologists that advance the science, go beyond their job descriptions. Everyone has some flexibility in their job that allows unconventional work. I would suggest that the new generations of fisheries scientists work at managing gamefish in the context of a broader ecological setting, which includes fish, reptiles and amphibians. It does not take much extra work to document the presence and general abundance of these species while sampling gamefish. In my later years, I came to the conclusion that most of the data that I was collecting was for someone else to use in the future. I think they would appreciate having a more complete database when they look back in time.

--

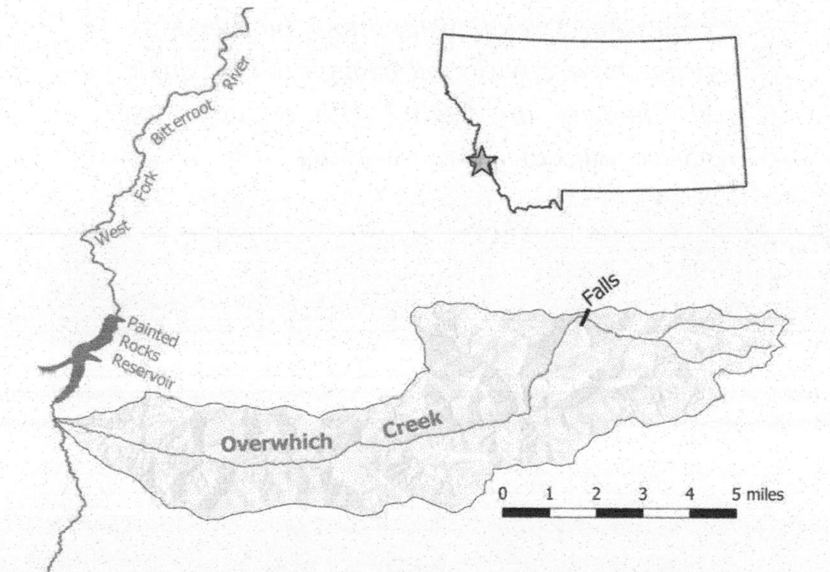

The location of Overwhich Creek in the West Fork Bitterroot River basin.

CHAPTER 3
Interpersonal Relationships & Partnerships

Interviews with
Travis Horton, Mike Jakober, Carter Kruse, and Brian Marotz

> *Relationships are difficult. Forming relationships that benefit fisheries can be even more difficult. I asked Travis Horton, Mike Jakober, and Carter Kruse to discuss the nuances of making and maintaining professional relationships and partnerships. Brian Marotz was kind enough to talk about his thoughts on dealing with setbacks and partnerships that aren't working.*

Travis Horton

Montana Fish, Wildlife & Parks – Bozeman*
MTAFS Chapter President 2013

*Travis now works as the environmental health director for the Gallatin County Health Department.

In addition to landowners, anglers, and other traditional stakeholders, I interact with many county, state and federal politicians. In general, I find it best to figure out some kind of common ground with difficult stakeholders. For example, talk to them about different folks that you may know in common. I often ask ranchers if they know people that I know and try to forge that common ground. Traditional landowners love to talk about weeds, so ask them "is this white top?" In my experience, a simple thing like asking about a weed and demonstrating some knowledge of a subject that landowners care about will pay dividends. Once they realize that you are like them in one way or another, the distrust level decreases. I recommend looking at "The Conservation Professionals Guide to Working With People" by Scott Bonar. He has a few great examples on finding common ground. Also, it is good to recognize and relate to their concerns, so they feel like they are being heard.

When trying to bring stakeholders to your side, one size does not fit all. The best story I recall is some vocal anglers wanting a biologist to stock a lake that had a unique strain of naturally reproducing rainbows. The biologist in this situation wanted to maintain the genetics of the strain and rely on wild reproduction. To address the criticism, he picked the most vocal critic and took him in the field. Together they collected gametes from the wild fish, hatched them and stocked the fry into the feeder stream. After this experience, the most vocal angler was defending the biologist to his friends and others that were critical of the management of the lake. Spend the time to develop relationships that are critical to your management program.

Working with stakeholders is critical, and I often advise my staff to approach situations dynamically. We all have our default communication styles: direct, overly scientific, passive, humble, etc. First be introspective, and identify your default communication style. Then learn how to read a situation, and employ a communication style that is going to best fit the situation.

--

Mike Jakober

U.S. Forest Service - Darby, retired
MTAFS 2017 Outstanding Professional Award

In my career, the most effective partnership I have developed is working with Montana Fish, Wildlife & Parks, particularly with Chris Clancy and Leslie Nyce. We have worked closely in the Bitterroot since the early 1990s, and over the years, our fish programs (Forest Service and state) have grown closer together, partly out of necessity (declining budgets and personnel), but also because we enjoy working together and are more effective when we work as a team. In recent years, the Bitterroot National Forest has also been growing partnerships with Trout Unlimited (both the local chapter and the national organization). The local TU chapter has funded a summer seasonal for 7 of the past 9 years, with the chapter picking up about half of the salary cost and the

National Forest contributing the other half. The national TU organization based in Missoula has also started surveying irrigation ditches for potential screen installations and is helping the Forest to fund the installations. The Bitterroot National Forest has also been able to replace approximately 15 fish-barrier culverts over the past 10 years because of partnership funding. These projects generally cost somewhere between $80K and $150K, so they aren't cheap. Our most effective partners have been the Ravalli County Resource Advisory Committee and FWP's Future Fisheries Program. From time to time, we have also been able to get some U.S. Fish and Wildlife Service money to install fish screens on irrigation ditches and modify diversion structures that are passage barriers to fish.

To form effective partnerships, I suggest starting small. Get to know your local counterparts in other agencies that are working on the same sorts of projects that you are. Once you get some small projects done together as partners, then you can start tackling bigger and more complex projects (e.g., our rotenone project in Overwhich Creek -- see Chapter 2). I also think partnerships are most effective when you team-up with someone who you like and enjoy working with, and someone in whom you share common values and goals. In other words, getting the interpersonal relational stuff figured out first is most important, then you can proceed to the actual biological issues. If you don't enjoy someone's company, you won't work together with them. Most partnerships in my career, I've needed to seek out, but a few have also fallen into my lap over the years. It is important to recognize opportunities and jump on them when you get the chance.

As state and federal budgets continue to decline (and I think that trend is inevitable because we live in a nation that is 34 trillion in debt and rising), and if you like getting things done on the job, all of us will have to work closer together to maximize our resources and get as much done as we possibly can.

--

Carter Kruse

Turner Enterprises, Inc. - Bozeman
MTAFS Chapter President 2009
MTAFS 2014 Outstanding Professional Award

Partnerships I've developed with state and federal resource management agencies have been the most effective in leading to successful natural resource outcomes. In particular, I've had good collaborative relationships and projects with Montana Fish Wildlife and Parks, New Mexico Department of Game and Fish, Nebraska Game and Parks, the U.S. Forest Service (Custer-Gallatin, Beaverhead-Deerlodge, Carson, and Gila National Forests's), and the U.S. Fish and Wildlife Service (Partners for Fish and Wildlife Program). These partnerships worked because there was mutual interest (shared objectives/goals), mutual trust (not immediately), and we each provided something the other(s) couldn't easily provide themselves. For example, our organization is often able to provide funding, personnel, logistics, and opportunity. Agencies are often able to bring the process, permitting, swagger, mandate, experience, as well as some funding and logistics . I've had successful partnerships with other private organizations too, like The Nature Conservancy, Trout Unlimited, etc., but typically our organization (Turner Enterprises, Inc.) and my projects have meshed better with resource agencies – in part because they are legally responsible for the resource.

Partnerships are based in relationships, at least the successful ones. I think they will be more successful if you care about one another personally. My advice to young biologists looking to develop partnerships is to first form internal partnerships. One of the first things you need to have is internal support within your agency or organization. Work with colleagues, supervisors, and administrators to line up internal support for outside partnerships. Only then should you develop or join external partnerships or collaborative efforts. A good basis for partnership is that you have mutual mandates or objectives/goals; common ground or interests; something to bring to the table; a willingness to work, and so on. Do not develop or join as a fundraising tactic, for self-importance, or just to be involved, and so on.

Turner Enterprises has been fortunate enough to leverage partnerships into some large restoration and conservation outcomes. For example, we've been involved in two of the largest successful native trout restoration efforts in North America. We accomplished the largest stream restoration projects in the Nebraska Sandhills. We manage the largest effort on behalf of Chiricahua leopard frogs in New Mexico. Lots of smaller projects benefit from partnerships too. I always say that we would get the work done by ourselves with time, but through partnerships the projects are better (better design, more intellectual power, etc.), bigger, and quicker; and likely more successful. And because of partnerships, our projects are often multi-jurisdictional or cross boundary. These are the fruits of partnerships.

Take your time, learn the landscape, and maximize your contribution to a partnership by prioritizing projects or partnerships that provide mutual benefit. Often, a specific project initiates a partnership, and if it goes well, leads to a broader partnership built out of hard work, trust, and successful outcomes. Develop personal relationships as you develop these organizational partnerships. Given good relationships, partnerships will be more cohesive and supportive for both parties. Further, be realistic about what you and your organization can provide, and then do your part. Under promise and over deliver. Bc honest and courteous. Don't hide your agenda. Don't abuse or use a partnership to further an agenda. Learn how to communicate effectively and learn about different communication strategies or types (e.g., personality testing). Respect older professionals.

To a certain extent, a private organization like ours needs the general support of state and federal resource agencies. At a minimum, we need permits from them to do much of our work. Thus, we have to seek out at least a basic partnership. The agencies themselves may not be as quick to reach out because they don't have to, at least in some circumstances. In my previous employ, I worked for a federal agency. I was a green rookie, but in public meetings I had power because everybody was looking to me as a federal employee to define the path forward. In my current position with Turner Enterprises, Inc., when I made the first overtures to state and federal agencies regarding proposed conservation actions, they all looked at me out of the corner of their eye wondering what I was really up to. As a green federal employee, I held the power, if not the trust of the folks around me. As a more experienced non-public employee later, I had no power or initial trust. But with hard work, and delivering on promises, the partnerships developed as a non-public employee are now much deeper and stronger.

As I've gotten older, I've gotten a bit more jaded on who really should be involved in a given project. Not everybody deserves a seat at the table. Only those committed, there for the right reasons, and willing and able to deliver should be maintained as partners. Too often, I think we measure our professional merit by the number of committees or partnerships or working groups we sit on or attend. It's not the number that matters, but the contribution given. We detract from a partnership or collaborative effort by not participating or contributing. We should all be able to differentiate when we are making a difference or not, and if not, butt-out or step-up. The biggest disruptions to collaborative efforts that I have been involved in are when a person or organization is not contributing – and personnel changes. Good personal and professional relationships run deep and are critical to good partnerships. When those are lost because of personnel turnover (e.g., promotions, leaving a job, retirement) partnerships can be disrupted or even become dysfunctional. New personalities have to be learned, new folks have to get up to speed, new professional opinions have to be reconciled, and so on. Turnover may be inevitable, but everybody should consider this dynamic and plan for these types of changes, by bringing new folks in early, rather than making abrupt changes, as well as committing to a project for as long as possible.

--

Brian Marotz

Montana Fish, Wildlife & Parks – Kalispell, retired
MTAFS 2015 Outstanding Professional Award

Setbacks were known to cause a blue haze of colorful expressions to waft from my office. Two shining examples leap to mind from my 34+ years with Montana Fish, Wildlife & Parks. I failed to close my office door during the first. In the early 1990s, I invested several years promoting a plan to retrofit Hungry Horse Dam to restore natural temperatures in about 45 miles of the Flathead River. Since the dam was completed in 1951, discharges from deep in the reservoir made the river ice cold when power was generated, and temperatures reverted to normal when discharges decreased. Our modeling of a new selective withdrawal system convinced Congress to appropriate funding to retrofit the dam. All was well in my world. So, imagine my rage when the U.S. Fish & Wildlife Service suddenly killed the project because they erroneously interpreted our data and concluded that Bull Trout would be harmed (someone down the hall complained about my screaming foul language). The situation launched me into damage control mode, and it took several weeks to correct the data misinterpretation and salvage the project. Selective withdrawal became operational in 1996, and the fishery responded exponentially, dramatically increasing the number of large Westslope Cutthroat and Bull Trout downstream. I learned that persistence pays.

A second memorable setback occurred after surviving a 5-year permitting process for a large Westslope Cutthroat conservation project in the South Fork Flathead River Basin when a governor-appointed FWP commissioner decided to kill the project (this time, I closed my door). To start at the beginning, in 1998 some friends and I caught what appeared to be hybrids (Rainbow x Cutthroat) deep in the Bob Marshall Wilderness, the largest stronghold for native westslope cutthroat trout in Montana. So we wondered, would the public allow us to eradicate non-native fish using rotenone in protected wilderness? Needless to say, the permitting process did not go smoothly. There were times when I wasn't sure I'd emerge from public meetings alive. Many small setbacks occurred, and again, persistence was required. But support for our controversial project grew with support from the Confederated Salish and Kootenai Tribes, Trout Unlimited, and the Forest Service, and eventually, the project was funded. Then, after those grueling 5 years, in walks a new fish & wildlife commissioner who told media outlets that he would stop this "insane" mountain lakes project, before ever speaking with anyone at FWP. That taught me that political appointees should fully understand a subject before publicly taking a position. Eventually, the media fire burned out because the other commissioners supported the project. But lots of unnecessary heartburn could have been avoided with a few phone calls.

Through it all, I learned that not all setbacks can be prevented. The only guarantee is that setbacks will occur, often unexpectedly and at the worst possible moment. When you hit a bump in the road, it's important to find the humor in these cosmic jokes. That said, communication is key, especially with your greatest critics. I urge you to actively seek out anyone who disagrees with you or could derail your effectiveness. Invite your detractors to express their viewpoints, and then listen carefully and envision their world view. Always be humble and remember that your greatest asset is your own personal credibility. Establish trust and maintain your integrity at all costs. Detractors often provide new gems of knowledge that will improve your actions. Armed with this knowledge, you can find common ground and devise more effective paths to your goals. One of the worst things you can do in public discourse is to speak down to people who might brand you as arrogant or a "book-learned idiot". If you can't answer a question, say you don't know and promise to provide your answer after further investigation, and be sure to follow through.

Name calling is great sport for many people - part of the human condition. Rather than take the bait and bite back, it's best to find the humor in the spectacle, but show respect. Jot down the clever names so you can relive the drama later and chuckle in private. I was once called a "Fish Nazi" and a "Picayune Government Hack", such treasures. I spent extra time with those folks. There were other times when I thought I might get punched in the face, but after further conversation, cool heads prevailed and some even volunteered their help.

Looking back, I learned a great deal from adversity. I was a bit brash in my younger days but mellowed with time. I'd like to think I've matured, but my friends might disagree. Heck, I might even disagree. When the heat was on, things weren't always enjoyable, and if I can exclude hypertension, the roughest points were positively stimulating. Better busy than bored, right?

--

CHAPTER 4
Losing Consensus on Flathead Lake

Interviews with
Jim Vashro and Barry Hansen

The events leading to the Flathead Lake fishery collapse are well-known in the small world of freshwater fish management; introduce mysis shrimp to increase kokanee numbers; inadvertently provide invasive lake trout with the perfect prey; watch kokanee, cutthroat, and bull trout numbers plummet. This is the story of two of the leaders who had to deal with the fallout.

I am immensely grateful to both Barry Hansen and Jim Vashro for providing their perspectives on what was (and still is) an extremely controversial and contentious topic. They may disagree on the management of Flathead Lake, but both obviously care greatly about Montana's fisheries.

Jim Vashro

Montana Fish, Wildlife & Parks – Kalispell, retired

Flathead Lake is the largest natural, freshwater lake west of the Mississippi River. The lake is influenced by three dams, and the water in the basin flows through a network of public and private lands managed by federal, state, provincial (Canada), tribal, and private organizations, three wilderness areas, and Glacier National Park. The U.S. Fish and Wildlife Service has jurisdiction over federally listed Bull Trout. All of this makes for a very complicated management situation.

There were 11 fish species native to Flathead Lake, and another 17 were introduced for food, sport, commerce and through illegal introductions. Fifteen non-native species persist and dominate the fishery. In an effort to grow larger kokanee, a non-native but popular sportfish, *Mysis relicta*, otherwise known as opossum shrimp or Mysis, were introduced into two lakes upstream of Flathead Lake in 1968. They drifted down, and Mysis were first detected in Flathead Lake in 1981, increasing to more than 250 per square meter by 1985. Too late it was discovered that Mysis were primarily being eaten by predacious Lake Trout and were actually competing with kokanee by feeding on the same zooplankton. Kokanee were wiped out, and the lake had been converted from a pelagic and nearshore fishery (kokanee, Westslope Cutthroat Trout, minnow species) to a benthic fishery dominated by Lake Trout and Lake Whitefish. Increased Lake Trout predation drove down numbers of native Bull and Westslope Cutthroat Trout. Further, before Mysis, Lake Trout had a reproduction bottleneck that kept them at low abundance but trophy size, and Flathead Lake was considered one of the top 10 Lake Trout fisheries in the US. What most people don’t recognize is that the pre-1980s Flathead Lake no longer exists, the fishery is profoundly altered. Until Mysis are eliminated, and no solution is known at this time, everything else is just treating symptoms.

The Flathead Reservation boundary is halfway up Flathead Lake. Under the Hellgate Treaty of 1855, the Confederated Salish and Kootenai Tribes (CSKT) were given the right to fish and hunt in all the usual and customary places. Due to the way mitigation money to offset losses from dams is allocated, FWP receives no mitigation funds for Flathead Lake management. This hinders what we can accomplish on our own.

Approaching 2000, the crash of kokanee and near crash of Bull and Westslope Cutthroat Trout had everyone concerned as we entered into a 10-year revision of the Flathead Fisheries Co-Management Plan. CSKT's stated goal was to return the fishery to conditions "pre-contact" (before European settlers). FWP has a dual mission – to protect and manage native fish species and to provide sport fishing. Both sides compromised with a draft Co-Management Plan that sought to rebalance the fishery by reducing Lake Trout through sport fishing to allow Bull and Westslope Cutthroat Trout to increase. A decision matrix was developed to adapt to changes in the fishery with changes in management strategies. Goals were set for annual fishing use in both the lake and river system.

Strategies included increasing angler access to the lake, liberalizing Lake Trout limits (to 100 fish per day), retaining a 30"-36" slot limit on Lake Trout to maintain a trophy component to both attract anglers and maintain large trout cannibalism on small Lake Trout), and fishing derbies to encourage angler harvest through random draws for prizes on all lake trout harvested and prizes on PIT-tagged Lake Trout. Through both natural cycles and fisheries management, Lake Trout numbers started to decline and Bull and Westslope numbers increased to about 70% of pre-Mysis numbers. However, changes were not fast enough for CSKT, and they advocated for large-scale agency Lake Trout gillnetting. This proposal was supported by groups concerned mostly with native fish restoration such as the Fish and Wildlife Service, Forest Service, Glacier National Park and Trout Unlimited (they fish mostly river trout). Netting was strongly opposed by most recreational lake anglers and charter captains, and FWP wanted to see the effects of regulation changes and fishing derbies before any major changes were made. Bull Trout were increasing, and Lake Trout now comprised most of the recreational fishery and charter fishery.

There are eight or so Lake Trout netting operations across the West aimed at reducing Lake Trout to benefit native trout. All will have to be continued in perpetuity since Lake Trout abundance will likely recover if netting is stopped. Netting is very expensive, costing $1 to $2 million annually in larger systems. Bycatch of native fish is a major concern. In Lake Pend Oreille, Idaho, a kokanee sport fishery has been re-established, but Bull Trout numbers have not increased due to bycatch of 400 or so Bull Trout each year. In Swan Lake, Montana, netting was suspended by FWP after nearly 10 years of removing 9,000 to 10,000 Lake Trout each year. Netting removed about 250 Bull Trout each year, and survival never increased to compensate for bycatch. Similarly, no other netting projects have shown a major recovery of native trout. Suppression netting is based on the collapse of Great Lakes Lake Trout after decades of commercial netting. But Great Lakes Lake Trout also faced commercial netting for other species (which sustained effort), predation by Sea Lamprey, chemical pollutants which inhibited reproduction and growth, and more than 160 aquatic invasive species. Western lakes have none of those.

FWP anticipated bycatch would prevent Bull Trout from increasing substantially in Flathead Lake. In addition, large Lake Trout were estimated to cannibalize around 400,000 small Lake Trout annually. Removal of large Lake Trout could lead to increasing overall Lake Trout abundance which could harm Bull Trout numbers. Lastly, netting would also remove large numbers of Lake Whitefish. After 1985 Mysis numbers had actually decreased instead of increasing as predicted. Predation on Mysis by Lake Whitefish and Lake Trout was considered a primary controlling factor. Removal of Lake Whitefish and Lake Trout could allow Mysis to increase. Increased Mysis predation could cause zooplankton to decrease which would allow phytoplankton to increase and potentially *affect water quality. In addition, the UM* BioStation theorized there are two stocks of Lake Trout in the lake based on the original Great Lakes stocks; deeper dwelling Lake Trout, known as "leans", eat primarily Mysis. To avoid bycatch of nearshore Bull and Cutthroat Trout and bycatch of deeper dwelling juvenile Bull Trout, netting to date has focused on deeper waters and larger mesh nets. More than 100,000 Lake Trout and substantial numbers of Lake Whitefish have been removed each year, catch rates of large Lake Trout have declined significantly but there has been no corresponding increase in Bull or Westslope Cutthroat Trout.

CSKT and FWP were working on a 10 year rewrite of the 2000 Flathead Lake and River Fisheries Co-Management Plan in 2010. CSKT wanted to move to aggressively net Lake Trout to remove as many as 143,000 per year. FWP cited decreasing Lake Trout numbers, increasing native fish numbers, and decreasing angler use on Flathead Lake as reasons to go more slowly. FWP also felt there hadn't been enough analysis on the plan to satisfy the Montana Environmental Protection Act (MEPA) and insufficient public participation to continue to participate. For example,

Lake Trout gillnetting was included as a strategy in the initial draft Co-Plan but withdrawn after very vocal public opposition. Lake Trout gillnetting was re-inserted when the final Co-Plan was published without public comment. The Co-Management Plan expired in 2010. Although many of the management strategies are still followed, FWP withdrew from the Co-Management Plan process. CSKT completed their own EIS and commenced Lake Trout gillnetting.

Coordination of fishing regulations also has suffered. CSKT removed the Lake Trout slot limit on the south half of the lake. FWP kept the slot limit to maintain both angler interest and lake trout cannibalism. FWP believes there are not enough trophy Lake Trout to impact native fish. Catch rates on large Lake Trout have declined since gillnetting was started. Bull Trout numbers have not increased substantially.

Unfortunately, monitoring and research on a body as large as Flathead is expensive and generally inadequate to remove uncertainty. There were also attempts to model the fishery. The modeling, research and monitoring were all subject to interpretation by various stakeholders depending on their goals and orientation. Although FWP was in agreement with CSKT on the bulk of the Co-Plan on subjects such as fishing access, habitat protection and monitoring strategies, there was disagreement on some overall fisheries goals and strategies. The most important tool now is monitoring to reveal the impact of ongoing management strategies for adaptive management.

Anglers and groups are very passionate about fish and fishing, I had serious attempts to get me fired 5 times over my career over disagreements on either habitat protection or fisheries management goals and strategies. It is important to practice active listening and have empathy for the viewpoint of other stakeholders. I commonly used advisory groups and met with stakeholders to encourage open communication.

Having solid science behind your decisions, sticking to the facts, and keeping your superiors informed (no surprises) will get you through most controversies.

However, the effort on Flathead Lake was not successful in that CSKT and FWP had different and somewhat conflicting goals for the fishery and are now pursuing different fisheries management strategies. As a sovereign nation, CSKT is able to chart its own course. FWP is subject to social and political pressure by anglers, constituents, and politicians that temper management decisions. There was substantial political pressure on FWP to conform to CSKT strategies, but FWP was steadfast that biology did not support agency involvement. FWP has remained open to another rewrite of the Flathead Co-Management Plan but as a total revamp with mutually agreeable goals, increased research, analysis and public input.

All the opinions stated above are mine and do not necessarily reflect management opinions and goals of Montana Fish, Wildlife and Parks.

--

Barry Hansen

Confederated Salish & Kootenai Tribes – Polson

In 2000, the Confederated Salish & Kootenai Tribes co-wrote a 10-year plan with Montana FWP to reset management of the recently disrupted Flathead Lake fishery. The re-write was contentious, because it meant someone would have to sacrifice, because the lake could not support overly abundant non-native Lake Trout without risking losing native fishes. CSKT's initial goal was full restoration of the native fishery, but we later compromised to accept a rebalancing that partially restored the native fishery and maintained status-quo sport fishing opportunity. It is important to understand that within broad sideboards set by the agencies, it was a citizen-advisory group that decided the direction and strategies for management in the plan. We thought at the conclusion of what had been an arduous planning process that the authors were in agreement with its direction. Soon after the plan was adopted, we co-wrote an article on the process called "Reaching Consensus", indicating some pleasure with our collaboration. At that time, we all seemed to be in agreement. We would later learn that the strongest disagreements came after the ink dried and implementation began. We had agreed to rebalance the fishery, but left ambiguity about the degree, timing and means of rebalancing (the devil is in the details).

While there were a wide range of stakeholders involved with planning, the strongest disputes were ultimately between the co-managers (CSKT and FWP). We held multiple public meetings and developed an interdisciplinary team with representatives from agencies, and public and non-governmental organizations. We tried to be open to all concerns and objectively address all comments.

During implementation, we found we disagreed about what we thought we had agreed to, and were therefore unable to implement the plan as a team. The target of much contention was the prospect of gillnetting as a tool of suppression. From the "Reaching Consensus" article, "the plan relies on recreational fishing to reduce nonnative fish numbers. If this strategy is not successful, more aggressive strategies such as commercial fishing or agency netting may be used to reduce nonnative fish". Within the first five years of the Co-plan's implementation, it was clear, based on extensive monitoring of Lake Trout abundance, that strategies more aggressive than angling were necessary to reduce the robust Lake Trout population. At this point we disagreed about something so fundamental as whether the Lake Trout population was increasing or decreasing. We made numerous efforts to resolve our differences, including third party reviews of our differing conclusions about monitoring and modeling data, as well as involvement in group dynamics and inter-personal mediation. We used independent boards and facilitators that included the Flathead Reservation Joint Fish and Wildlife Board, the Montana Fish & Wildlife Commission, the Northwest Power and Conservation Council, multiple assembled scientific panels including the Bonneville Power Administration's prestigious Independent Scientific Review Panel, and even private facilitated sessions. Reviewers were in nearly unanimous concurrence that native species recovery was important, Lake Trout were not in decline, and more aggressive suppression was necessary to recover native fishes. Finally, after nine years of delay, the Tribes pushed on with what they considered was the original compromise made under the plan.

While the Tribes sincerely worked to accommodate all possible stakeholders, it should be noted that CSKT's primary responsibility is to the tribal membership. The Tribal vision for the Lake-River system was much more native fish oriented than the compromise agreed to in 2000 with the Co-Plan. Therefore, the Tribes had very little latitude to further compromise on the subject of non-native Lake Trout vs. native trout in Flathead Lake.

I think the foundational issue of dispute was based on values, which created an intractable situation that affected judgements and the analysis of facts. The fundamental questions are: 1) Are native fish and biodiversity more valuable than sport fishing, 2) Are they mutually exclusive?, and 3) Could sport fishers accept any sacrifice to ensure survival of native fish? The Tribes and many others concluded that without some sacrifice by sport anglers, there would likely be a total sacrifice of native trout. Ultimately, I think disagreements in values became disagreements over science. The Tribes position, generally corroborated by all the reviewers, was that exclusion of native trout by Lake Trout was inevitable without intervention, although as with many issues in fisheries science, we recognized some uncertainty existed. That bit of uncertainty opened the door for copious counterarguments.

The opposing scientific argument was that the system had adapted to Lake Trout and reached a new equilibrium, so there was no need to suppress Lake Trout and jeopardize the opposing value of fishing opportunity. I interpreted this scientific difference to really be the result of the underlying disagreement over values. When we have a value to protect, we are most receptive to arguments and interpretations that support that value. Therefore, in this case it was necessary for independent third parties to review the data without the confirmation bias of either co-manager. To further complicate matters, fishing guides tried to sway opinions by fabricating data to advance their argument that Lake Trout abundance had declined, when we had no empirical evidence to support their case.

I believe this case study likely provides an example of when not to change your thinking despite some very loud stakeholders. Were there credible science, well expressed social or political reasons to change course, I would. I came away from the process thinking that, while agencies must be transparent, objective, and open to stakeholder ideas, that openness shouldn't interfere with an agency's obligation to achieve its mandate even if there is substantial resistance. Decision-makers must be willing to make hard decisions and ensure that compromise is real, meaning that achieving opposing goals is feasible, not just empty words. I think there are cases where it is appropriate for management agencies to follow conservation goals even when those goals may not be the most popular among some in the public.

--

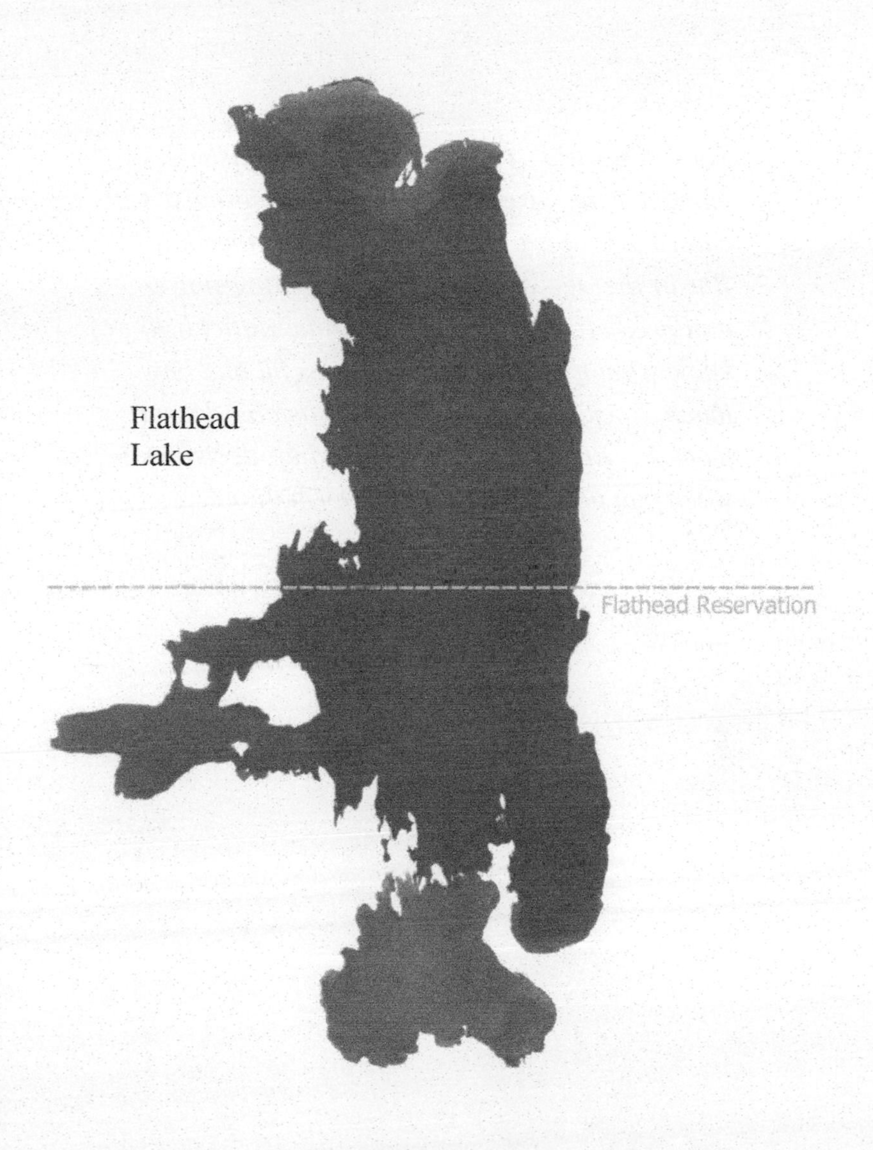

CHAPTER 5
Data & Uncertainty

Interviews with
Wade Fredenberg, Chris Hunter, and Jim Dunnigan

The scientific underpinnings of fisheries management are what set a biologist's knowledge apart from the angler's anecdote. But in the age of unlimited computing power and predictive modeling, it can be difficult to know what type of data is most useful and how much stock to put in the conclusions drawn from it. And what about making decisions when you only have limited information?

Wade Fredenberg

U.S. Fish and Wildlife Service – Kalispell, retired
MTAFS 1991 Outstanding Professional Award

When making management decisions, the best information is quantitative empirical data that is generated in testing a hypothesis. I consider the type of information that Dick Vincent used to support the decision to stop stocking streams as kind of the classic – i.e., pre- and post- change population estimates with a control. However, that quality of information is seldom available, so most times we rely on case histories from similar management scenarios. In those situations, one must account for the similarity of the circumstances; are the same species complexes and similar environmental conditions present? I would also rely heavily on the redundancy of responses or the circumstantial evidence. A good example is the effect of Lake Trout on Bull Trout populations. When you have a dozen or more case histories in different water bodies that replicate the same outcome (i.e., Lake Trout dominate and Bull Trout decline to remnant status), then the management decision can be fairly certain.

These kinds of circumstances often apply to fishing regulations. However, in those cases we seldom have direct estimates of mortality rates and other parameters, so the management decisions are often instituted on a trial basis, with the implication that we can always go back if the new approach doesn't work. What frequently complicates the determination is that these decisions are seldom purely biological, typically incorporating a huge social element. A case-in-point is the argument over barbless hooks. Case after case has shown that, scientifically, there's little to no population-level benefit of barbless hooks. Yet the institution of barbless-hook regulations is still a hot topic and widely embraced by a major portion of the public.

Often, the most difficult management decisions to institute are those that involve liberalizing, rather than restricting regulations or harvest. It's often perceived that, "it must be working, so why mess with it?" even though there's little or no scientific support. In my experience on the Bighorn River, despite our agency policy encouraging anglers to harvest Brown Trout, very few chose to do so. In fact, in practice, management changes are typically viewed with extreme skepticism and often become more dependent on social science than biology. Because we are asked to make decisions about more places and circumstances than we can reasonably collect data, deciding where, when, and how often to collect it is important. Ideally, if a management change is instituted, there's a pre-change monitoring program that will allow us to assess the efficacy of the change. In practicality, that data often doesn't exist, or if it does, there are existential factors (e.g., environmental stochasticity) that masks any clear interpretation of the result. The best monitoring is that which is consistent, repeatable, simple, and easy to describe and interpret. That's why standardized population estimates, spawner escapement counts, redd counts, juvenile-abundance indices, and the like, are the most proven and reliable monitoring methods.

An example of this is the Bull Trout redd count database conducted on adfluvial spawning tributaries in the Flathead. It checks all those previously described boxes and is so easy to interpret that the public was literally trained to await the results published in an annual news release, as regularly as Thanksgiving. However, as the database has extended into its 5th decade (40+ years), it has also become apparent that counting conditions are more variable than first acknowledged, and the redd counts don't necessarily translate into an accurate index of year-class strength. Useful as it is, there's a tendency to become overly reliant. It's like a check-engine light in that it may effectively warn that something is wrong, but it won't necessarily tell you what the problem(s) are, and the solutions can be very controversial (e.g., the decades-long dispute over Lake Trout suppression in Flathead Lake – see Chapter 4).

It's important, especially with emerging technology, that we strive to identify and implement new monitoring techniques. Examples are the recent ability to track and tag fish and even more recent applicability of eDNA and other genetic techniques. The challenge here is to integrate these techniques into current monitoring programs. Ideally, cheaper and simpler monitoring methods will supplant older more labor-intensive methods (e.g., mark-recapture population estimates). A caveat is that it often takes a change of personnel to facilitate a switch from one method to another. As biologists, we all reflect the era in which we "grew up" and tend to rely on those techniques. In transitioning from one monitoring method to another, the ideal scenario is that there's a few years of overlap so that the new method can be verified.

The best biologists in every era, the folks we all look up to, are those who made good "gut level" decisions and were able to eventually prove them right with follow-up monitoring that showed a positive response. The classic, of course, was the story of Dick Vincent and wild trout. But there were others. Joe Huston deserves recognition in Region 1 for being an early and passionate advocate of native fish in Hungry Horse Reservoir, such that the entire South Fork Flathead is now recognized as a native fish refugium.

Decisions to introduce new species are often irreversible and frequently have been the single biggest complicating factor in future management. The decision to introduce Mysis in lakes in northwest Montana was a gut-level decision that was made in Region 1 that was pushed by popular opinion. But George Holton [former assistant fisheries division chief] later called approving it in Helena the biggest mistake of his career. To me, the most important point is that some seemingly innocuous decisions can have extremely serious long-term consequences, even into perpetuity. The structured approach introduced through the Montana and National Environmental Policy Acts has been helpful in reducing the number of poor choices made by managers, but sadly, the public has assumed a bigger role in illegally introducing species, both intentionally (Northern Pike, Walleye, Smallmouth Bass, Black Crappie) and inadvertently (zebra/quagga mussels, alien plants, etc.). Increasingly, due to unintended ecosystem consequences and potential impacts on dwindling native species, there is less room for error, so more scrutiny and a higher standard for potentially irreversible management decisions will be a requirement for the future.

Regarding the greatest challenges with data collection and management, the *Big 2* have probably not changed very much throughout the 75-years or so that modern fisheries management has been in existence. First, is the challenge of adequate resources. Very seldom does a manager have the luxury of adequate staffing and equipment to collect and analyze all the data one would like to have in facing a major management dilemma. As a practical matter, we have to cut corners and make assumptions, and if those assumptions are wrong, they can lead to poor decisions. It will likely always be that way. One can make up some of that ground by communicating within the fisheries community and networking with our peers. I have always been a huge advocate of communications within the internal fishery community, but also externally. It is especially important to network with those outside our own agencies, be they in other organizations or other states and provinces. I'm a big fan of expert panels so long as those in power are willing to receive the messages conveyed.

The second enduring challenge is simply organizing and interpreting the information. I'm a supporter of the recent push to input field data into permanent databases that can be searched and analyzed. One of the biggest challenges to my generation of biologists, [which I viewed as kind of the F2 generation (1970's to early 2000's), following the pioneer generation (1940's through 1960's)], was to be able to easily consult information that was generated previously. Prior to the 1980's, most reports were not digitally available, and the data itself was even more inaccessible. Even worse, due to the time crunch and relative difficulty of processing data (the pioneers didn't even have calculators for God's sake, let alone computers), a treasure trove of field data simply sat in files, eventually to be boxed and often thrown out. With modern capabilities, this should no longer happen. Dan Isaak and others have even illustrated through their Climate Shield project that there are unlimited possibilities through the use of crowd-sourced information.

A third challenge I see, and this may be a more recent phenomenon, is in maintaining a connection between the rapidly emerging, model-driven and highly technical, empirical biosphere and the grassroots observation-based science that our profession was founded on. I doubt that I, as an older biologist, am alone in finding that it sometimes seems like half of the journal articles I read employ mathematical models and interpretation that are out of my league. As a result, I'm often incapable of reading these articles and evaluating their merit. The concern I have is that, too often, those studies are published by academics who may not know the difference between an anode and a cathode. I am not arguing that these types of articles shouldn't be published, but rather that we must continually strive to ensure highly technical findings are supported by real-time observations. It is vitally important that our discipline maintain that contact with the resource that leads to good gut-level decisions, despite what the models might say. I learned more in the field (and sometimes over beers) about the resources I was responsible for than I ever did in the office.

Lastly, all management decisions do not have the same consequences. Simple angling regulations, for example, may be relatively easy to reverse and leave no lasting consequences. Management decisions that can alter native species complexes (e.g., species introductions) are much weightier. The more lasting the consequences of a decision, the more likely it needs to be supported by some pretty good data. The precautionary principle ("First, Do No Harm") comes into play here.

--

Chris Hunter

Montana Fish, Wildlife & Parks – Helena, retired
MTAFS Chapter President 1991
MTAFS 1987 Outstanding Professional Award

In making decisions, the data I found most useful was long-term trend information. Most of our data does not meet standards of strict confidence intervals, so I felt most comfortable with the long-term trend information. Depending on the question being asked, I also relied on presence-absence data. In some circumstances, fish health and genetic information were also very important. In retrospect, I would like to have had more ecological data; I think most agency data collection is directed at popular game species. As a result, we miss important ecological trends or changes. It would be great to have long-term information on key indicator species-both game and nongame. I would include herps, macroinvertebrates, and habitat if I could. Of course, staff and funding are always limiting factors. Ultimately, you have to balance the questions you think you will be trying to answer with staffing and funding limitations.

When we had little data, I would tend to make decisions conservatively. For instance, when we considered stocking Pallid Sturgeon in an attempt to recover that population, I was very conservative about bringing in fish from federal hatcheries in the Dakotas when I did not have good fish-health information for the receiving water. Follow the 'do no harm' principle.

There are also many issues related to data collection and management. Standardization is a big one. Are the crews collecting the data the same way across the state and over time? Similarly, is the data being analyzed and managed the same way? Is the data being analyzed and utilized in decision making? The preparation of annual reports tends to become rote. It is seldom the highest priority for biologists. Consequently, data may not be analyzed and incorporated into the reports. Often, they do not take the time to compare the current data to long term trends.

Something else to consider are the opinions of stakeholders and other managers. People working in the field can feel tremendous pressure from local opinion leaders, economic interests and politicians which can influence decision making. If an issue is elevated to a citizen board or commission, the role of stakeholders can be elevated. At this point, it is critical that the agency have good data in which they have confidence. The staff must also be able to present the data in a clear and understandable manner for both the commission and the public. Having good data is critical to the integrity of the biologists, management and the agency. If the decision-making body does not have confidence in agency data collection and analysis, it really hurts the agency and the resource. In this situation, anecdotal info from stakeholders can trump any data presented. Stories can be more compelling than data. Ultimately, we are a political society. Decisions are made based upon public opinion. The role of good data is to try to inform public opinion.

One last thought. Whether in a public meeting or a fish and game commission meeting, an anecdotal story by an angler or two can be more influential than a ton of data. Someone told me once to try to make presentations into a story that includes data. So when talking about Paddlefish populations and regulations, talk about how long they have been on the planet, how they used to make huge migrations like salmon, how they use their paddle for balance and feeding, and weave the data into that story. This can be harder to do, but it is actually fun and, I think, more effective.

--

Jim Dunnigan

Montana Fish, Wildlife & Parks - Libby
MTAFS Chapter President 2021

Most fisheries management decisions basically boil down to actions intended to influence either the abundance or size structure of the population in question. The three population dynamic rate functions that collectively influence the abundance and size structure of a fish population are recruitment, growth, and mortality. A solid understanding of which of these rate functions may be limiting the population from attaining a desired management objective is critical to successful fisheries management. However, estimation of these rates is often challenging, causing managers to defer to measuring the desired outcomes of abundance or size structure.

Life is full of choices throughout our adult lives, including our professional careers. At the early stages of my professional career, I was rather surprised at the flexibility I had in my first professional position. I quickly learned that I had remarkable latitude in determining the types of data I would collect. A previous supervisor gave me some sound advice. He said: If you're doing X that means you can't do Y. Of course, what he really meant was that it isn't possible or practical to monitor everything under the sun due to time, funding, or FTE constraints, and you have to determine what's most important. Here are a few thoughts about things I've learned on the topic after working as a fish biologist for 26 years in state and tribal government.

The first big decision that most professionals probably make without even really thinking about it is which species to monitor. Identifying monitoring priorities often includes value judgements that are frequently ingrained within the agency or entity that employs you. The reality is, depending on where your paycheck is coming from, much of our effort is devoted to game and threatened & endangered species and the habitat that supports them. Before collecting any data, you really need to ask yourself the purpose. If you can't explain to yourself the ultimate use and purpose, I would suggest you pause until you can. I think most monitoring can be divided into three general categories. 1) Trend-Status monitoring (sometimes called asset inventory) data are collected for the purpose of determining the distribution and/or population trajectory for a particular species. These types of monitoring data are often collected to determine if management actions are warranted to conserve or restore the species. 2) Effectiveness monitoring is performed after a management action is taken to determine if that action had the intended biological response. 3) Applied research is what I like to term research that will likely be used to inform management alternatives. Before you can implement a management action to solve a problem, you first have to understand the underlying cause(s) of the problem. Applied research often involves collection of various biotic and abiotic information to understand the physical and biological relationships that influence the population dynamics of the species of interest. Informed management alternatives can be developed that will likely result in an intended biological response.

A common issue I've seen throughout my career related to effectiveness monitoring of management actions is an inadequate pre-management dataset where too few or sparse data makes statistically valid pre and post comparisons ambiguous. Standardize data collection methods to the greatest extent possible to ensure consistency and maintain data quality through time, and document those protocols. This is especially important as staffing changes through time. Think through any analyses involving the data before you collect it. If you have a fundamental understanding of future analyses, it will ensure you are able to use the data for the intended purpose.

When making decisions with little to no data, the first rule of thumb is *do no harm,* and *if it's not broken, don't fix it*. Making management decisions with little or no data essentially means we are making decisions based on opinion, and we all know that just about everyone interested in the resources we manage has opinions. Make sure you have a reasonable understanding of the biological processes influencing the population-of-interest before you start tinkering. We achieve this by developing hypotheses and collecting data to test these hypotheses. Making management decisions with little or no data is tantamount to trial-and-error guesswork that risks losing credibility with constituents, squandering public resources or resulting in biological consequences contrary to your intentions.

Most young professionals are fortunate to directly participate in much of the field data collection efforts. However, the natural professional progression often includes promotion to levels with increased administrative responsibilities that reduces field time. This often necessitates the supervision of field personnel that perform much of the field data collection. Never ask someone else to do something you wouldn't consider doing yourself. Have faith in your field workers but explain the importance of the data being collected and what it will be used for, even if you are in the field with your technicians. If you are a field worker collecting the data, ask your supervisor these questions if he/she hasn't explained it to you. Developing this knowledge builds ownership and teamwork that ultimately results in higher data quality.

--

CHAPTER 6
The Removal of Milltown Dam

Interview with
David Brooks

The construction of Milltown Dam was completed in 1908 on the Clark Fork River, just below the confluence with the Blackfoot River. Later that year, a flood washed tons of heavily contaminated sediments into the newly completed reservoir, the result of substantial waste deposits from the upstream mining operations at Anaconda and Butte.

In 1981, the reservoir was listed by the Environmental Protection Agency as a Superfund site - a designation indicating that substantial, long-term cleanup was needed. In 1992, the Superfund designation was expanded to include much of the Clark Fork River upstream of the reservoir. On March 28th, 2008, hundreds of people gathered to watch Milltown Dam be breached.

David Brooks, now director of Montana Trout Unlimited, completed his dissertation on the history of Milltown Dam in 2012 at the University of Montana and published a book on the matter in 2015. I asked him to relate what he had learned about the science and types of data that brought the dam down.

David Brooks

Montana Trout Unlimited – Missoula

Many, many scientists and skills were involved in all the stages of identifying the problems and coming to the conclusion to remove the Milltown Dam. One of my favorite stories about the Milltown Dam Superfund site is from right after it was designated a Superfund site in 1992. Two University of Montana geology/hydrology professors skied out onto the reservoir while it was frozen with a handful of grad students. They used a chainsaw to cut holes in the ice, then lower a pole down to the bottom of the reservoir to take samples of the sediment near the dam. This investigation helped determine that the cause of previously identified arsenic contamination of drinking water in the area was coming from the sediment behind the dam, which in turn had come from the smelting of ore in Anaconda and Butte decades earlier. Other organizations included the Missoula County Health Department which was involved in water quality sampling, determining the threat Milltown posed to Missoula's water system, and downstream river health. Montana Fish, Wildlife & Parks handled lots of the fisheries issues, as did biologists from the U.S. Fish and Wildlife Service. The Montana Department of Environmental Quality and the U.S. Environmental Protection Agency oversaw most of the science on how to remediate the toxic sediment and, ultimately, how to handle dam removal and restoration. Removal of the dam, and restoration, was part of the early boom in the dam-removal industry. Toxic waste removal, transport, and storage, as well as how to restore a river channel and riparian area, were all technical skills on display at Milltown. The data collection and science that ultimately led to the removal of Milltown Dam was truly a team effort.

From a fisheries perspective, some of the most important data was collected by FWP biologist David Schmetterling who began studying the impacts the dam was having on migrating fish in the 1990's. The theory was that fish were no longer trying to migrate up past the dam since it had been in place for nearly 100 years. Dave immediately began documenting tens of thousands of fish pilling up below the dam trying to get past it and follow their evolutionary instinct to migrate up to spawning waters in the upper Clark Fork River, the Blackfoot River, and their tributaries. Dave found native Bull Trout, Westslope Cutthroat Trout, and Mountain Whitefish, as well as others. So, he started catching and tagging Westslope Cutthroat then releasing them above the dam to see where they went. Lots of area schools 'adopted' one of these tagged trout and monitored the fish's spawning migration. When one Cutthroat that the Seeley Lake school had adopted got eaten by an invasive Northern Pike in the Milltown Reservoir on its journey back downriver, students, parents, and the public got pretty fired up about the harm the dam and reservoir were doing to native fish. Dave's work helped establish and promote the knowledge that the dam was seriously impacting fisheries by impeding migration and, hence, spawning. Around the same time, 1996, U.S. Secretary of the Interior Bruce Babbitt showed up along the shores of the Blackfoot to announce the listing of Bull Trout under the Endangered Species Act. The dam's impact on an ESA trout amplified the public's concern about the dam and the local fishery. Those two things made the health of the Blackfoot River a much bigger part of the story of the dam's impacts.

Some of the most important people skills that were behind dam removal were outreach, education, &

marketing. It's hard to get the public fired up and keep them engaged in a 30-year process during which the goal - dam removal - wasn't even part of the discussion until halfway through that process. So, the ability to take lots of technical information and lots of litigation and lots of interagency bureaucracy and distill it for the public in such a way that people were willing to pay attention, willing to make phone calls, write letters, or sign postcards to advocate for dam removal and restoration was huge. Montana Trout Unlimited and the Clark Fork Coalition were at the center of those public-outreach efforts. With the help of professional marketers, they got the message out that Milltown was a toxic time bomb, that a free-flowing river was better than a reservoir full of arsenic, and, ultimately, the campaign motto stuck: "Remove the Dam, Restore the River." Part of the reason that campaign was so effective is that it was fun and did more than just remove bad stuff. It was about making the river better, freeing it. People really embraced the idea of making positive change, not just preventing something bad.

However, patience was key at Milltown. Had a solution been decided on quickly after the site was designated by Superfund, it would have involved leaving the dam in place and replacing the local drinking water system. Because of delays, and redoing the environmental impacts of different solutions many times, the option of dam removal was evaluated more thoroughly and gained support. Critics of Superfund often say that cleanups take too long. The truth is, at Milltown, if it hadn't taken so long, this poster child for Superfund cleanup would not have happened, or would not have come out so well.

In short, the Milltown Dam removal, remediation, and restoration is still considered one of the most successful Superfund cleanup sites, even by the EPA, which administers Superfund. Out of Milltown came the practice of engaging the public early on, getting local stakeholders involved, being creative about solutions and the Three Rs of Superfund: Remediation, Restoration, Redevelopment. That is the watermark of Superfund sites and the best example is Milltown. The Redevelopment portion of that equation is still unfolding. The new Milltown State Park opened in June 2018. There's a new Kettlehouse brewery and amphitheater in Milltown-Bonner alongside a portion of the Blackfoot that was part of the old mill. Trails now lead to and through the old reservoir site. Obviously the technical skills involved were many and very important. But those skills would not have been put to use, especially the restoration stuff, without the overwhelming public support. Getting people involved was the number one skill that shaped the fate of Milltown.

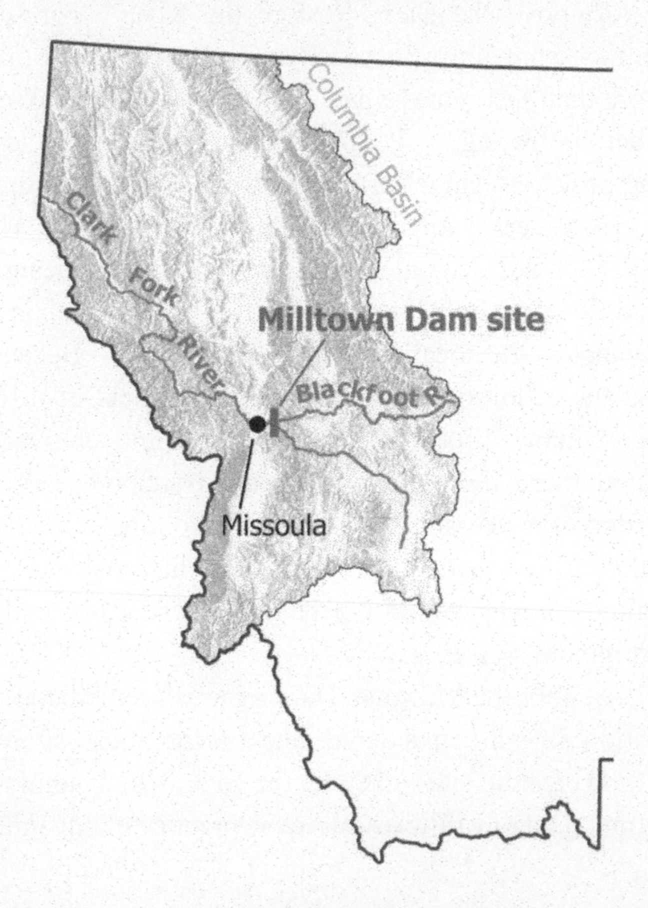

CHAPTER 7
Fisheries Management in the Future

Interviews with

Caleb Bollman, Tom McMahon, Leslie Nyce, and Pat Saffel

As of the writing of this book, Montana can proudly boast that not a single native species of fish has been extirpated from the state. It is in no small part due to the efforts of dedicated fisheries conservationists that this is true. I asked Caleb Bollman, Tom McMahon, Leslie Nyce, and Pat Saffel for their views on the future of Montana's fisheries and what challenges lie ahead.

Caleb Bollman

Montana Fish, Wildlife & Parks – Miles City

I think the future of fisheries in Montana is bright. There are a diversity of habitats in the state, large prairie rivers, mountain streams, high mountain lakes, large reservoirs, and smaller impoundments that are home to a variety of species. Montana's fisheries are important to residents and non-residents alike for their many values including recreational, consumptive, ecological, and aesthetic. A solid foundation from over a century of fisheries management in the state has set up current fisheries professionals for success in expanding our scientific knowledge base and in providing decision makers with information for the crucial resource-management decisions they make. There are plenty of challenges facing fisheries in Montana including curbing the spread of aquatic invasive species, ensuring adequate water quality and quantity in the face of changing climate, erratic weather, and high allocation demand, maintaining genetic purity in species and diversity in populations, to name a few. However, the cultural value of fisheries and widespread concern for their health and well-being makes me confident that these challenges can be opportunities for Montana to step up to the challenge.

Identifying future problems early and working with the landowners, public representatives, community organizations, and other resource agencies toward collaborative solutions is important. Gathering sound scientific information that maintains our credibility, and provides a basis for conversation and decision making, is essential. Continuing to promote the value of fisheries and connect people to the resource will build advocacy and ensure, when problems arise, people care enough to find solutions.

What can you, as a professional, do to secure a more positive future for the fishery resource? I think we can do our part to secure a positive future for the fishery resource by working hard, seeking continual improvement, listening carefully, continuing to learn, and being flexible, honest, and persistent. I also think it is more important than ever to look for what connects us rather than what divides us, and build on those connections in looking toward long-term, collective results and improvement rather than short-term individual benefits.

--

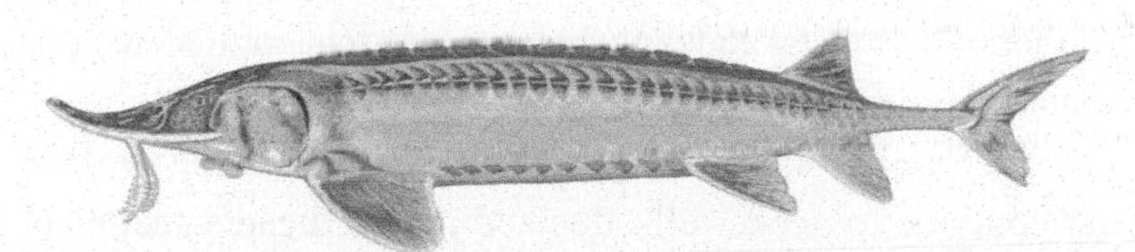

Shovelnose Sturgeon (*Scaphirhynchus platorynchus*)

Tom McMahon

Montana State University – Bozeman, retired
MTAFS Chapter President 1999
MTAFS 2001 Outstanding Professional Award

I am cautiously optimistic regarding the future of Montana's fisheries. Climate change is looming, and all the unpredictability associated with the various scenarios—greater disease outbreaks, less water, expanding invasive species, increased physiological stress, and changing species distributions— will definitely strain our creativity and resourcefulness as resource managers in the future. But at the same time, I am optimistic because we have a strong conservation ethic in Montana, we have strong institutions and a regulatory framework in place that has a history of conserving resources—some of the best in the nation. We have a strong university system that wants to help find answers to resource problems, and we have an increasing number of interested and engaged citizens that want to see our valuable resources protected.

To secure the most positive possible future, we will first continue to need well-educated and dedicated technical experts who provide the best available scientific knowledge to help guide us as a society in making the best decisions. So, we have to continue to support our universities and our resource management agencies in gathering information, conducting studies, and in explaining to the public the likely consequences of alternative actions during decision making about our valuable resources in Montana and elsewhere. We, as a citizenry, also have to respect and listen to resource managers and scientists during our deliberations. Along these lines, we need an engaged, open-minded citizenry that isn't afraid to experiment with different outcomes in search of the best solutions to difficult issues. The blooming of citizen science is certainly a relatively new, positive feature that should really help with facing future challenges.

As a professor, my main roles have been education—educating the next generation of fisheries scientists and resource managers—and research—learning about natural systems so that we have an ever-improving knowledge base for making informed resource-management decisions. Both of these activities are important, I think, in securing a more positive future for fishery resources. As an educator, you try to give students the tools to be ready for the future, and to do that, I think students need to be armed with an understanding of how we gain reliable knowledge, as well as a broad understanding of how ecosystems operate and how they respond to various management actions. They also require knowledge of the various tools used to gather and analyze ecological information, and coupled with an understanding of some conservation history, how we got where we are today and why. I think this blend of technical knowledge, of learning how ecosystems function and how management decisions are made, and a sense of history, are all important in sparking students' interest and passion and laying the foundation of the scientific and ecological understanding that will assist them in developing their own conservation ethic.

Two experiences as an educator give me a lot of hope for the future. When I see smart, hardworking, creative undergraduate and graduate students gain passion and knowledge and go on to obtain influential positions in resource management agencies, that gives me a feeling that the future is in good hands. I also have had the interesting experience teaching courses to K-12 science teachers from all over the country, and I have been impressed with their creativity and dedication to teaching young students about science and critical thinking, nature, and conservation issues—that gives me great hope as well.

--

Leslie Nyce

Montana Fish, Wildlife & Parks – Hamilton
MTAFS Chapter President 2017

I am hopeful that Montana's amazing fisheries resources will be here for future generations to enjoy despite some tough challenges (a shifting climate and invasive species are two examples). Resident and non-resident individuals seem to value the fisheries across the state, many recognize just how special these resources are, and ideally this translates into actions that will maintain and protect the fishery. That may look like safeguarding this incredible public resource, protecting stream/riverbanks from development, or not fishing as much during warmer summers. I believe that the various entities managing Montana's fish (federal, state, etc.) strive to provide opportunities for people to enjoy these resources while at the same time, protecting, preserving, maintaining, and enhancing the fisheries and the habitats necessary for survival.

If you are a young professional, be optimistic. Think about the difference you as an individual can make to contribute to the fisheries resources where you are working. Don't be afraid to think outside the box, maybe implementing new methods to get the job done. This does not mean ignoring the past. Learn from it. Can it be improved? Collaboration is huge. Work with others to protect the resources that you cherish.

Women working in the world of fisheries have come a long way since the early 2000's. If a young woman wants to work in the fisheries world, I would tell her, "do not give up; pursue that dream!" Talk to other women in the fish world, learn from their stories, and learn from their experiences. Be prepared for what may be hard, physically demanding work, and be ready, willing, and able to work with a lot of men – which does not necessarily mean that they will not respect you as an equal. It will be tough since you may be the only woman in the room or on a field crew, but you can do it. Believe in yourself!

Lastly, communication and respect are key for any relationship to work and especially the biologist-technician relationship. These two positions often work together daily, and you'll spend A LOT of time together. Be flexible and willing to learn from each other.

--

Pat Saffel

Montana Fish, Wildlife & Parks – Missoula
MTAFS Chapter President 2015
MTAFS 2011 Outstanding Professional Award

There are two ways to view the future of fisheries in Montana. On the positive side, I think there is growing appreciation for outdoor recreation, including fishing. With more people taking part in fishing and other aquatic recreation, there is greater support for these resources. However, not all people recreating around water have an understanding of the connection of fish to water and their particular pursuits. Constant awareness is needed. On the negative side, there is less water and more demand for it. Lower snowpack, earlier snowmelt, population growth and expansion is affecting water supply and quality as well as habitat. Climate change will likely adjust our fisheries. Trying to preserve native fishes and valuable cold-water fisheries will be difficult in many cases. Water, the basic element of aquatic habitat, is likely to be the most important conservation focus.

One of the most important things we can do is locally manage water as a public resource more than as an individual right that is enforced by the users. Water allocation to natural resources may have to be sacrificed in some areas to maintain priority resources. We need to increase awareness of where our water comes from and where we can be more efficient in order to maintain the quality we currently enjoy. Habitat protection, including water, is about enhancing awareness and appreciation for natural resources so it is something people will want to do, rather than are regulated to do.

We can chip away at improving our understanding and management of fishery resources by advancing science. However, it's hard to argue that we need a lot more understanding that fish need water. As a professional, I work on issues in the geographic and social scope where I have influence. I have worked on water awareness campaigns, water rights for instream flows, native and cold-water fish habitat, and population enhancements and protection. It is encouraging that people still want fishery resources, and we as a society are investing in them. Still, the management of climate and water may be different as it affects lifestyle changes that many are not willing to compromise.

--

CHAPTER 8
The Musselshell Watershed Coalition

Interview with
Mike Ruggles

Interacting with landowners, ranchers, and farmers is a big part of a fisheries biologist's job, especially in parts of the state dominated by private land. Conservation districts and watershed groups often act as a primary point of contact between stakeholders and biologists. This is a case study about the Musselshell Watershed Coalition, an extremely effective watershed group.

Mike Ruggles

Montana Fish, Wildlife & Parks - Billings
Great Plains Fishery Workers Association President 2006
MTAFS 2016 Outstanding Professional Award

The origin of the Musselshell Watershed Coalition (MWC) in central Montana started from a point of contention regarding water availability. A drought led many sections of the river to dry up in the 1980's. Forced to come together by the state Department of Natural Resources and Conservation, water users came to realize the power of working together. In 2009, the MWC was created to formally organize grant proposals, project coordination, and work on other issues. Ken Frazer and Jim Darling from Fish, Wildlife & Parks were present at the start. Their efforts set the stage for future conservation work. Today, water use is measured. This ensures legal use and minimizes illegal water withdrawal, allowing for water to be in the river more. Senior water rights in the lower river are being met despite the cycle of drought.

As water management changed, thanks in large part to the coordination of the MWC, it opened up discussion about what the fishery had been and what it could be. Fisheries had been studied by FWP's Bill Wiedenheft in the 1970's and early 1980's and by Wade Fredenberg in the early and mid-1980's. This gave us some baseline information about barrier locations and fish-movement connection to the Musselshell for at least the first 120 miles upstream of the Missouri River, particularly for Channel Catfish and Sauger. A major benefit of the MWC for the basin was opening up lines of communication between private landowners, and state and federal agencies. This reaches up to legislation in Montana and Federal support.

In 2011, a large flood caused a lot of hardship in the Musselshell. Roads flooded, people were cutoff from getting groceries and drinking water, dams were flanked, and channel abandonment caused the river to become 10% shorter. The partnerships formed by the MWC were critical in the rebuilding efforts and recognition that we could build water infrastructure back that was better for everyone.

After discussions with MWC members about the rebuilding process, they recognized fish passage was important and wanted to get additional funds and support for the Deadman's Basin diversion dam repair. We set the target species for passage as Northern Redbelly Dace, a species-of-concern and a fish with more challenges than most for swimming speed, burst speed, and jumping abilities. Designs that meet criteria for the dace met most other fish present. The new diversion and fish passage was completed in 2016. I believe the relationship-building through MWC has led to a greater appreciation of native fishes like Burbot, Sauger, Channel Catfish, and others.

Beyond keeping more water in the river, the MWC has helped coordinate flood analysis and floodplain improvement, improve fire resiliency, complete willow/soft bank stabilization, channel reconnection, and barrier removal or fish-passage projects. The partnerships resulted in more funding for many projects on water quality work, infrastructure repair, abandoned mine clean up, and removing floodplain berms.

There are lessons from the MWC that I think can likely be applied elsewhere. Foster a space for diverse needs and ideas in a way that allows discussions to occur. Provide opportunity for people to discuss problems and potential. As government biologists, we work for the people. Our personal ideologies don't always mesh with the realities on the ground. If you fight this, it will be difficult to make progress and be included in important discussions. Be open to hearing desires and stories from those that know the area. Share challenges to meet objectives and provide space for others to share. Help look for alternatives that are mutually acceptable. Build plans together and share the work. Strive to understand other perspectives, and be understanding and often sympathetic to those ideas. Learn from mistakes, acknowledge and share those, and move forward.

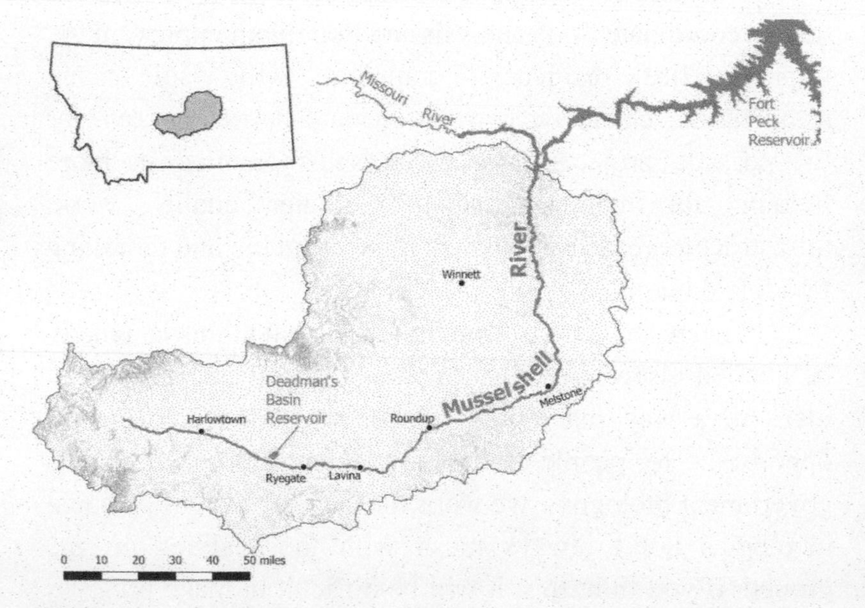

The location of the Musselshell River basin in central Montana.

Fish passage on the Musselshell River at Deadman's Basin Diversion. The passage is designed to pass large and small species, alike.
Courtesy of Musselshell Watershed Coalition

CHAPTER 9
Concluding Notes

Niall Clancy

Here, I attempt to summarize some of the most poignant observations and suggestions, focusing especially on those that were mentioned by multiple interviewees.

Effective Biology through Affective Biology

I believe the preceding chapters make clear that the most effective biologists are affective biologists - the ones who you can tell really care. In the words of the late Brad Shepard, "the first thing a person must have is the personal commitment to become the best professional fish scientist or worker they can become." When setbacks occurred, those that managed to get things done were simply those who were the most determined. Those included in this book are surely some of the fisheries professionals most committed to conservation, management, education, and continual improvement in the state. Almost all were, or are, active members of the American Fisheries Society.

A Gender-Biased Sample

One thing that stuck out to me in the course of preparing this work was the startling lack of female professionals in the field. At the time of this writing, only 2 of the 26 FWP management biologists, and 1 of 7 regional fisheries managers are women. While some of this disparity can likely be explained by greater participation of males in recreational fishing (an issue in and of itself), I don't believe it fully explains these lopsided figures. A brief look at university ecology programs will show a far more gender-balanced set of interested students. Researchers that study hiring practices have noted that people in charge of hiring are often unconsciously looking for candidates that remind them of themselves. Even when gender is expressly forbidden to be considered in hiring decisions, I suspect that hiring committees made primarily of men tend to prefer skills that they themselves possess or value. While I don't have a clear solution, it is something our profession needs to remedy. Otherwise, we risk excluding different perspectives that would ultimately lead to better management and conservation outcomes. For young women considering a career in fisheries, Leslie Nyce suggests, "talk to other women in the fish world, learn from their stories, and learn from their experiences."

It All Starts With Personal Relationships

Most interviewees commented on the importance of working well with those inside and outside your particular agency or group. Of note, showing humility, having respect for the legitimate needs of stakeholders, and not "taking the bait and biting back," as Brian Marotz said, will provide the best shot at success. Both Carter Kruse and Mike Jakober state that effective partnerships that accomplish big things often start with good personal relationships between a few individuals that then tackle larger and larger problems. According to Wade Fredenberg, Chris Clancy, and Amber Steed, one key to forming these relationships is making sure not to fall into the "we know best" trap and being open to learning from people in other agencies or organizations.

Have Someone Else Make Your Case

Local resource agencies are often the most trusted source on a controversial topic, but that doesn't always mean the public will be on your side. As Travis Horton points out, convincing a "vocal critic" to make your case can be a winning tactic. Chris Hunter also suggests that providing an "anecdotal story [from] an angler or two, can be more influential than a ton of data."

Compromise Where You Can, But Where You Can't, Don't

While maintaining a professional demeanor at all times is paramount, there will be stakeholders that flat out disagree with you. Pat Clancey, Jim Vashro, Barry Hansen, Brian Marotz, and David Brooks all discussed issues that are amongst the most controversial in recent state fisheries decisions. When you know in your core your cause is just, when, as Aldo Leopold says, "a thing is right when it tends to preserve the integrity, stability and beauty of the biotic community," stick to your guns. In the cases of Cherry and Overwhich Creeks, the South Fork Flathead, and Milltown Dam, this dogged persistence paid off. And, even in light of little compromise at Flathead Lake, Barry Hansen maintains, "there are cases where it is appropriate for management agencies to follow conservation goals even when those goals may not be the most popular among some in the public."

First, Do No Harm

I was struck by how many times this saying, *paraphrased from the Hippocratic oath*, was repeated. Bob Gresswell, Wade Fredenberg, Chris Hunter, and Jim Dunnigan all mentioned it. The meaning attributed to it was twofold. First, to conserve ecosystems, one must not allow substantial harm to befall them, and a good biologist does what they can to prevent outside forces from doing permanent damage to the waters in their purview. The tools at your disposal to prevent this harm depend on the agency or organization you work for but often include the issuance (or non-issuance) of permits. The power to issue these permits is regularly used to modify otherwise harmful projects so that harm to the fishery is minimized while still allowing necessary work to move forward. Second, when a management decision is needed but you have incomplete information about how different actions might influence the fishery, the "do no harm" or conservative approach is often best. Wade Fredenberg suggests that the permanency of the potential action should be carefully considered; if tweaking angling regulations, most changes can be reversed if further evidence suggests it was not the right move. But with actions that may be longer lasting, such as Chris Hunter's example of deciding from where to gather Pallid Sturgeon broodstock, making a conservative choice that has the least chance of negative consequences is often best.

Have One Foot in the Field and One Behind the Computer

There's no doubt that the computer age has led to the rapid adoption of more complex forms of statistical modeling than any one of us can possibly understand. Brad Shepard suggests that you "gain statistical knowledge so you can analyze your data and talk to high-powered statisticians, as necessary." Wade Fredenberg and Beth Gardner both mention that knowing how data is collected in the field is vital to a complete understanding of these new tools, and Matt Boyer urges us to cultivate our naturalist skills so that we can merge ideas from multiple disciplines, including statistics. Jim Dunnigan brings up the issue that fisheries sample sizes, no matter how painstakingly attained, may not be large enough to determine if a population is changing. Considering this will often be important.

The Future is ... Warm

All contributors to the chapter on the future of Montana's fisheries, Caleb Bollman, Tom McMahon, Leslie Nyce, and Pat Saffel, were optimistic about the future biologists of the state. All four also talked about their concern regarding the impacts of climate change. While we are still learning exactly how, where, and what species will be most affected, Pat Saffel quips that, "it's hard to argue that we need a lot more understanding that fish need water." Efforts to cooperatively manage the amount of water in streams, including effective management of instream flow rights, will be increasingly critical for many species.

Remember the Ecological Perspective

Lastly, Chris Hunter and Chris Clancy both urge us to think more broadly about the species we manage. Both acknowledge that most agencies are under-funded for monitoring or managing nongame species, but as Chris Clancy says, "younger professionals should broaden the gamefish culture of their agency. We took ecology classes for a reason. The focus of efforts will continue to be gamefish, but with little extra work, agencies can collect data on the fish that may not be pretty, grow large, take a fly, or taste good." Growing concern about the status of nongame fishes like Northern Redbelly Dace, Pearl Dace, Plains Sucker, Sicklefin and Sturgeon Chub, and Columbia Slimy Sculpin - or even other nongame aquatics like Rocky Mountain tailed frogs and western pearlshell mussels - illustrates the urgency of this broadened perspective. It is up to the current and future fisheries professionals of the state to implement this change.

Westslope Cutthroat Trout (*Oncorhynchus lewisi*)